THE INTERCONNECTED UNIVERSE

Conceptual Foundations of
Transdisciplinary Unified Theory

THE INTERCONNECTED UNIVERSE

Conceptual Foundations of Transdisciplinary Unified Theory

Ervin Laszlo

Foreword by Arne Naess

Afterword by Karl Pribram

World Scientific
Singapore • New Jersey • London • Hong Kong

Published by

World Scientific Publishing Co. Pte. Ltd.
P O Box 128, Farrer Road, Singapore 912805
USA office: Suite 1B, 1060 Main Street, River Edge, NJ 07661
UK office: 57 Shelton Street, Covent Garden, London WC2H 9HE

British Library Cataloguing-in-Publication Data
A catalogue record for this book is available from the British Library.

First published 1995
Reprinted 1999

THE INTERCONNECTED UNIVERSE

Copyright © 1995 by World Scientific Publishing Co. Pte. Ltd.

All rights reserved. This book, or parts thereof, may not be reproduced in any form or by any means, electronic or mechanical, including photocopying, recording or any information storage and retrieval system now known or to be invented, without written permission from the Publisher.

For photocopying of material in this volume, please pay a copying fee through the Copyright Clearance Center, Inc., 222 Rosewood Drive, Danvers, MA 01923, USA. In this case permission to photocopy is not required from the publisher.

ISBN 981-02-2202-5

Printed in Singapore.

Foreword

by Arne Naess

Now, at the end of the century, it appears that at its beginning something happened that will have a profound influence far into the future. Without losing touch with observation under carefully registered conditions, the level of abstract thought in natural science theorizing increased immensely. The steps of deduction increased proportionally. The fundamental theoretical postulates of the new physical theories tend to contain concepts more and more abstruse, seemingly of an arbitrary character. That which for most people must appear as an increase of unfounded speculation, and more like invention than discovery, is in reality a consequence of a liberating synthesis of thought and imagination.

The wave of development impinged on philosophy in unpredictable ways. The creative work of Ervin Laszlo is a brilliant testimony of how conceptual imagination—deductively related to careful observation—can make us see the cosmos, and our place within the cosmos, in new ways that are of great inspirational value. Reality as conceived by Ervin Laszlo has what I call 'gestalt character'—a predominance of internal rather than external relations.

Ervin Laszlo often uses the term 'speculation' and does not pretend that, for instance, recent research *implies* a definite conception of consciousness. He differs from the more uncritical author Stanislav Grof, who says that modern consciousness research *reveals* "that our psyche has no real boundaries." The present book is admirably sober— no spiritual *Schwermerei* in the Kantian sense.

Laszlo rightly emphasizes the simplicity of the theories he explains so well. It is a simplicity that takes a great deal of training to perceive; beginners may find the theories 'complicated'. Yet they have a remarkable simplicity in relation to their great power of explanation—in regard to the new, formidable range of phenomena covered by their conceptual framework.

Some philosophers, and I am one of them, intensely admire the creations of mathematical physics—the little that we understand of quantum physics and of Einstein's theory of spacetime and its cosmological consequences. One of the things we wonder about is what would have happened if the famous universal constants, such as Planck's constant, had different values, perhaps only a little bigger or smaller than what they do have. Physicists tend to answer that the evolution of the cosmos would have been vastly different; the cosmos may not even have been created. But what could explain the existence of these preciously tuned constants? Did they evolve out of something preexisting? But how? Ervin Laszlo attempts to answer, even if tentatively, questions at this level of depth.

In regard to these questions it does not seem possible that any kind of observation could be of any use—at least not today. For the present answers must be speculative, as Laszlo admits. Many postulates are required, many premises must be tentatively adopted so that they may explain otherwise unexplainable phenomena. Yet comparatively simple and basic reflections can help us to formulate the relevant questions with clarity.

"A universe of isolated components governed by pure chance could not give rise to the orders that now meet our eye," Laszlo writes. I would say that even if this were conceivable, in such a universe the next moment might see the destruction of everything. If so, why bother to contemplate it?

Obviously we not only can, but should, bother to contemplate a universe that is systematically interconnected. Laszlo formulates the question his book seeks to answer: "How could the evolutionary trajectories of the many things we observe, or infer from observations, not only diverge, but also converge with each other, even when they are separated in space and time?" (p. 4)

A central part of Laszlo's conceptual framework is the quantum/vacuum interaction (QVI) hypothesis. This is a highly sophisticated theory, rather than a hypothesis, in my terminology. I hope that it will be taken seriously by the few professionals who can fully judge its scientific merit. Other readers of the book are advised to joyfully continue reading even if some concepts (for example, 'the Dirac sea of the quantum vacuum', 'the sequential emergence of diversified yet ordered systems', or 'interconnections that bias evolutionary trajectories') are not entirely understood by them; they can get many 'aha-experiences' nevertheless.

FOREWORD

Greatly simplified, one might say that Laszlo envisions a world that is constantly created, and where every event that happens locally, even an event in one's consciousness, is connected with events that happen everywhere else. There is no absolute determinism in this world, although evolution exhibits a measure of bias in favor of certain trajectories and directions, rather than others. Complete randomness is out of the question or, strictly speaking, incomprehensible. The time is ripe to work out exact theories of general evolution in the cosmos, transcending the usual distinction between theories of physics and biology, and concepts of the physical and the mental.

There are many of us in science and philosophy who wish to see a growing trend of bold theory formulation inspired by such courageous yet unpretentious efforts as the present study by Ervin Laszlo.

CONTENTS

Foreword *by Arnes Naess*   v

Preface   xiii

PART ONE: THEORETICAL CONSIDERATIONS

1. A Transdisciplinary Framework for Evolution   3
   - Space and time connections in nature   5
     - *Interconnections in evolution*   5
     - *The relevance of fields*   9
     - *Temporal field connections*   11
     - *The evolutionary functions of a holographic field*   14
   - The fifth field   15
     - *Physical roots of the fifth field*   16

2. Basic Concepts of Quantum/Vacuum Interaction (QVI)   25
   - The vacuum postulates   27
     - *Quanta as solitons*   28
     - *Secondary waves in the vacuum*   30
   - The interaction postulates   33
     - *Vacuum-quantum interaction*   33
     - *Micro- and macro-scale applications*   34
   - Related hypotheses   37
     - *Bohm's implicate order*   38
     - *Stapp's Heisenberg quantum universe*   39
     - *Sheldrake's formative causation*   42
     - *Summary*   44

PART TWO: EMPIRICAL EXPLORATIONS

3. QVI in Microphysics and Cosmology — 49
   - Interaction effects in microphysics — 50
     - *The quantum-interpretation decision-tree* — 51
     - *An interpretation of quantum nonlocality and coherence* — 53
   - Interaction effects in cosmology — 60
     - *The problem of the constants* — 60
     - *Non-Big-Bang cosmologies* — 63
     - *The multicyclic QVI scenario* — 66

4. QVI in Biology — 73
   - Problems of the synthetic theory — 74
   - Interaction effects on the organism — 79
     - *The in-formation of the organism* — 79

5. QVI in the Cognitive Sciences — 87
   - The interactive varieties of human experience — 88
     - *ESP* — 88
     - *Anomalous recall* — 93
   - The interpretation of interative experience — 95
     - *Near-death and past-life experiences revisited* — 97
     - *A fresh look at ESP* — 103

PART THREE: SUPPLEMENTARY STUDIES

(i) QVI Dynamics in the Brain — 113
   by Attila Grandpierre

(ii) Philosophical Perspectives of QVI Cosmology — 119
   by Mauro Ceruti

(iii) New Concepts of Space and Time — 125

(iv) Creativity, Archetypes, and the Collective Unconscious — 129

(v) Problems and Prospects of Transdisciplinary Unified Theory — 137

Afterword *by Karl Pribram*	143
Bibliography	147
Index of Names	157
Index of Subjects	161

PREFACE

Conveying an understanding of the world around us that is rigorous and detailed, as well as complete and consistent, is a perennial ambition of both science and philosophy. In the past, satisfying this ideal required considerable leaps of imagination, binding together the observed facts with rationalistic metaphysics, if not with myths and mysticism. In recent years the prospects for progress have improved. The empirical sciences now research ever more domains of the observable or observationally inferable world, and they do so with increasing rigor. Even more significantly, the increasingly rigorous probings of nature come up with significant similarities: basically analogous processes are discovered in independently pursued programs and fields of investigation. Not many years ago these were still thought to be mainly superficial — suggestive analogies without necessarily deeper meaning. Today, however, theories that unify entire domains of research are actively sought in most fields, first and foremost in the physical and the biological sciences. Grand unified theories are a legitimate, and often even a dominant, program in physics, while the elaboration of the synthetic theory fulfills a parallel function in biology.

The separation *between* the physical and the life sciences has not been satisfactorily bridged, however. Scepticism regarding the adequacy of laws of physics for explaining the laws and regularities of life persists, and the postulation of quasi-biological laws in the physical realm is viewed with distrust. Though in the light of past experiences such attitudes are clearly justified, there are methods of unification available today that can do significantly better. General laws and regularities are discovered in field after field, and they neither inflate physical processes into biology, nor reduce biological processes to physics. Attempts to bridge disciplinary gaps are now approaching a new threshold of validity and cogency.

The next step in creating rigorous as well as reasonably complete and consistent transdisciplinary theories calls for explicating the basic process whereby the universe evolves in an unbroken (though not necessarily linear) sweep from basic physical entities to the complex open systems that are physicochemical, biophysical, biochemical, biological, ecological, and even psychological in nature. The theory that describes these developmental processes is unified theory of the transdisciplinary variety. Unlike unified theories in the life sciences and grand-unified theories in physics, it does not remain limited to the one or the other disciplinary field but encompasses physics and biology, and ultimately the human and the social sciences, as special elements within a general scheme. Such a scheme, to be scientifically acceptable, must do better than the insightful but speculative attempts of traditional philosophers. Its conceptual framework must be optimally rigorous and consistent, yet simple and parsimonious, and its empirical implications need to be consistent with findings in both the physical and the life sciences. The view put forward here is that the theory that satisfies these requirements is *transdisciplinary unified theory*.

Unless we assume that physical and living nature are inseparably disjoined in reality, we have to acknowledge that there is no genuine unification in science or philosophy that is not transdisciplinary in scope. Indeed, there is no genuine 'grand' unified theory that is not a transdisciplinary unified theory—a genuine GUT is always a TUT.

Fully elaborated TUTs are as yet in the future; the quantitative analysis that a mature theory presupposes is yet to be developed. It is not likely to be developed, however, in the absence of the clarification of the fundamental notions on which that theory would rest. These basic conceptual elements can be already outlined—doing so is the aim of this study. The scheme offered here suggests alternative and highly unified ways of conceptualizing known and as yet anomalous data; exploring ways anomalous data can be rendered consistent with known data; and discovering the basic concepts that could integrate the known as well as the previously anomalous data into a meaningful general theory.

In the context of this endeavor it is well to remember that science grows not only through the piecemeal refinement of established tenets, but through the proposal of new tenets that go beyond the scope of the old. Evidently, not all new tenets become established theory; but those that do, make for basic innovation in the theories and concepts of science. It is in the expectation that the present study contains the seeds of such an innovation that it is now placed in the hands of the reader for assessment and elaboration.

ACKNOWLEDGMENTS

The ideas presented here are the result of a long period of research and reflection, interspersed with intense and insightful discussion with a considerable number of friends and colleagues. I cannot thank all of them in the space of this note, but would like to take this opportunity to express my appreciation to Mauro Ceruti, in whose series the basic concept has first seen publication (*l'Ipotesi del Campo* Ψ, Bergamo 1987)—and who also contributed the note on cosmology in Part Three—; likewise to Jean Staune, who worked with me in preparing a further elaborated version in France (*Aux Racines de l'Univers*, Paris 1992) and to Christopher Moore, who helped me prepare the still more developed version in England (*The Creative Cosmos*, Edinburgh 1993). In researching the text now before the reader, I have benefited in particular from in-depth discussions with Karl Pribram—whom I also wish to thank for contributing the Afterword—, Ilya Prigogine, John Wheeler, Henry Stapp, Ignazio Masulli, Attila Grandpierre (who contributed the section on dynamics in the brain), Mária Sági, Stanley Krippner, Roberto Fondi, Karan Singh, Mario Varvoglis, Christine Hardy, Bernhard Haisch, Errol Harris, Peter Saunders, Allan Combs, and Henning Bråten. I have received important comments and contributions from Arne Naess—to whom I owe the here published Foreword—, John White, Søren Brier, David Lorimer, C.J.S. Clarke, David Chalmers, R.J. Gilson, Willis Harman, Carl Upton, Donald Keys, Paul LaViolette, Ib Ravn, Michael Conforti, David Peat, Ignazio Licata, Min Jiayin, and Luo Huisheng. Nobody but I am responsible for the use I have made of their insights and materials, including any misunderstandings and flaws that may have cropped up in my interpretation.

December 1994

PART ONE

THEORETICAL CONSIDERATIONS

Chapter One

A TRANSDISCIPLINARY FRAMEWORK FOR EVOLUTION

The universe, as far as we know, evolved from an initial state of *chaos* (in the Greek sense of unordered complexity) to the present state of *cosmos* (or ordered complexity). This ancient insight is reaffirmed in the contemporary sciences without, however, fully and satisfactorily answering the problems implied by it.

We now have reason to believe that all that exists in our immediately observable universe is the product of a process of structuration that began in an explosive instability fifteen (or perhaps only seven or eight) billion years ago. Life, and even mind, arose in the course of this process. If so, there is a question of evolution by design versus evolution by chance—or by chance versus necessity. Evolution by design presupposes a Designer, and that supposition is beyond the bounds of science. The action of final causes is likewise unacceptable: the process is better conceived as generating its own goals than having them provided ready-made. But the problem of chance versus necessity is not as easily laid to rest. As the heating up of this time-honored argument shows, it is by no means resolved. A few basic considerations can be stated, however.

The process that gave rise to the spectacle that now greets our eyes cannot have been governed by chance alone. It also cannot have evolved as it did in a fragmented manner, with its elements detached from one another. A universe of isolated components governed by chance could not give rise to the orders that we now find; indeed, pure chance could not exist in the universe even if it coexisted with strands of order. If a series of pure chance events had punctuated the developmental process, the systems that emerge out of that process would diverge randomly among themselves. Unless reality was governed by something like Leibniz's pre-established harmony, higher levels of order could not grow out of a set of randomly diverging lower-level orders—given a process that is subject to pure chance, even previously ordered systems would each grow their own way. This is manifestly not the case, even if chaos often surrounds the crystallizations of order.

The question we face is this: how could the evolutionary trajectories of the many things we observe, or infer from observations—cells and people, as well as atoms and molecules, stars and galaxies, species and ecologies—not only diverge but also converge with each other, even when separated in space and time? In a chance-riddled universe the enormously ordered systems that underlie the phenomena of life and mind are unlikely to come about; yet life, mind, and the many layers of ordered complexity in the cosmos and in the human world did come about. It is reasonable to conclude that pure chance did not (and does not) dominate the evolutionary process: there must also be a significant degree of non-randomness. This, however, does not require necessity of the classical deterministic kind. It suggests merely one of several possible varieties of correlation among the evolving systems.

When a set of parts or elements are constantly and systematically interconnected, the set generates order spontaneously. One element in-forms the others: the system as a whole becomes self-referential. Pure chance cannot exist in such a system; full randomness presupposes the disconnection of the random event from its past, as well as from its milieu. But an adequately interconnected system may well allow a degree of freedom in the selection of the evolutionary trajectory of its parts. The crucial factor is the type of causality that is exemplified in the interconnection. This need be neither of the classical linear variety, where an event A fully determines an event B, nor of a global kind, where the full set of events $A, B, \ldots n$ determines each event A. Connections must be selective, though not exclusive of other connections; they must be of the kind where each event is preferentially correlated with a specific set of other events. The system's preferential correlations—they exemplify nature's algorithms—determine the types of order that evolves. If they are sufficiently specific yet broad-ranging, the order that evolves does not merely repeat the order that has already emerged, but creates possibilities for attaining genuine novelty within the range of possibility.

Given the ordered complexity that meets our eye, the reasonable assumption is that, somehow, preferential interconnections must exist in nature. If so, there must be some factor in the universe that interconnects (and therewith nondeterministically correlates) the evolving systems. Finding this factor is not a simple matter. The difficulty is that, though it has surrendered necessity of the deterministic variety, mainstream science suffers from the opposite flaw: under certain conditions it allows disconnections among systems and environments, and affirms that some events may be governed by chance alone. Yet, if the considerations put forward here are not mistaken, processes that are free of some variety of correlation both with other events and entities, and with the past of the events and entities that manifest them, cannot produce the overall developmental system that would be capable of generating the observed orders. This suggests, in turn, that the received laws and entities of contemporary science may need to be completed with an optimally rigorous and economical hypothesis that can account for constant interconnections among systems— interconnections that give rise to universal yet preferential correlations among them.

It is the objective of this study to outline such a hypothesis. In this opening Chapter we marshal the arguments and assemble the conceptual framework for this task.

Space- and time-connections in nature

Interconnections in evolution

In nature, elementary particles build into atoms and atoms build into molecules and crystals. Molecules in turn build into macromolecules and into still more complex cellular structures associated with life; and ultimately cells build into multicellular organisms and these again into social groups and ecologies. It is not necessary, nor indeed reasonable, that each of these processes of assembly should obey categorically distinct laws. The same basic laws, functioning as nature's algorithms, could create the interactive dynamics whereby complexity builds in the universe from the level of particles to that of organisms, and then to that of the ecologies and societies of organisms.

Until the last few decades, theories of universal evolution were produced by philosophers who complemented the lacunae of scientific knowledge with speculative insight. Although speculative, such works as Bergson's *Creative Evolution*, Herbert

Spencer's *First Principles,* Samuel Alexander's *Space, Time and Deity,* Pierre Teilhard de Chardin's *Phenomenon of Man,* and Alfred North Whitehead's *Process and Reality* stand as enduring milestones of evolutionary thinking. Recently, however, concepts and theories have been developed that promise to lift evolution as a universal phenomenon from the realm of philosophical speculation into that of scientific investigation. Among these theories nonequilibrium thermodynamics—the thermodynamics of irreversible processes— merits special attention.

Classical thermodynamics was concerned with the transformation of free energy into waste heat in closed systems, with the consequent breakdown of order into randomness. In nineteenth century physics, the ultimate implication of this line of thought was the heat-death of the whole universe. But in the first half of the twentieth century, physicists have been exploring new approaches. Lars Onsager's 1931 study 'Reciprocal Relations in Irreversible Processes' was pointing in the direction of irreversible processes that move systems away from, rather than toward, thermodynamic equilibrium. In 1947 Ilya Prigogine devoted his doctoral dissertation to the behavior of systems far from equilibrium, and in the early 1960s Aharon Katchalsky and P.F. Curran elaborated the mathematical basis of the new science of nonequilibrium thermodynamics. These investigators showed that by concentrating on gradual changes in closed systems, classical thermodynamics has failed to confront real world systems—nonequilibrium systems that evolve nonlinearly and are open to energy flows in their environment. Such systems are basic to life: as Schrödinger noted in mid-century, 'life feeds on negentropy.'

An open system far from thermodynamic equilibrium dissipates entropy as it performs work: in Prigogine's terms, it imports free energy from its surroundings and exports entropy to its environment. If such a system imports more negentropy than the entropy it dissipates, it grows and evolves. In open systems change in entropy is defined by the equation $dS = d_iS + d_eS$, where dS is the total change of entropy in the system, d_iS the entropy change produced by irreversible processes within it, and d_eS the entropy transported across the system's boundaries. By contrast in an isolated system dS is always positive, for it is uniquely determined by d_iS, which necessarily grows as the system performs work. However, in an open system d_eS can offset the entropy produced within the system and may even exceed it. Thus dS in an open system need not be positive: it can be zero or negative. An open system can be in a stationary state $(dS = 0)$, or it can grow and complexify $(dS < 0)$. Entropy change is then given by the equation

$d_eS = (d_iS \leq 0)$, which means that the entropy produced by irreversible processes within the system is shifted into its environment.[1]

Evolution—the negentropic complexification of a system—is triggered when a critical fluctuation pushes a far-from-equilibrium system still further from thermal and chemical equilibrium. The new order arises in the interplay of critical fluctuations during the crucial phase-change of an instability. If the system is to evolve rather than devolve, at least one out of the many possible fluctuations must 'nucleate'—that is, diffuse rapidly throughout the system. If and when it does, the whole system undergoes a bifurcation: its evolutionary trajectory forks off into a new mode. The dynamic regime it then assumes defines the norm around which the system's typical values will thereafter fluctuate.

The evolutionary process as a whole depends critically on randomness in the system. "Only when a system behaves in a sufficiently random way may the difference between past and future, and therefore irreversibility, enter into its description," wrote Prigogine with Isabelle Stengers in *Order out of Chaos*.[2] "The 'historical' path along which the system evolves as the control parameter grows is characterized by a succession of stable regions, where deterministic laws dominate, and of unstable ones, near the bifurcation points, where the system can 'choose' between or among more than one possible future. Both the deterministic character of the kinetic equations whereby the set of possible states and their respective stability can be calculated, and the random fluctuations 'choosing' between or among the states around bifurcation points, are inextricably connected. This mixture of necessity and chance constitutes the history of the system."[3]

Perturbations, the random interplay of critical fluctuations, and the bifurcation that follows upon the nucleation of some of the fluctuations, are the key elements that define the interactive dynamics responsible for the evolution of far-from-equilibrium systems in nature. The question that needs to be posed is whether this dynamics can provide a full and satisfactory explanation of the order that arises in nature.

Let us reexamine the proposition. In Prigogine's dissipative systems dynamics the specific developmental path of the evolving system is the prey of chance: the bifurcation process is described by stochastic equations that allow a spread of probabilities in the outcome. Neither the past history of a system, nor the flows that reach it from its environment, decides which of the possibly numerous fluctuations will actually nucleate. This, however, creates a difficulty. If neither the past nor the environment determines the outcome of a bifurcation, then the dynamic regime that arises in a complex system is at the mercy of a random selection from among the many fluctuations that arises in the system. The way evolution unfolds in *one* system becomes fully unpredictable; and the way it unfolds in *many* systems is likely to be diverse. If systems in nature were driven by the Prigoginian dynamics, they would tend to diverge and diversify. Even if two systems were to start in the same state and with the same initial conditions, they would still diverge in the course of their evolution as each system came to be exposed to a different set of external influences and different patterns of internal fluctuation.

Prigogine spoke with good reason of a 'divergence property' basic to the evolutionary process. But he would also need to speak of a 'convergence property,' for systems not only diversify in space and time, they also converge within higher-level systems. Without the latter property the universe would be populated with nothing more interesting than an array of chaotically varied and mutually uncoordinated particles, randomly colliding in hydrogen and helium gases.

If theory is to be adequate to the facts, in addition to the dynamics of divergence, it must also describe the dynamics of convergence. The conceptual basis for that dynamics is furnished by the constant interconnection of the nonequilibrium system with its environment. Prigogine generally affirms the interconnection of such systems with their milieu— essentially, the rest of the universe—but suspends it during the period of bifurcation: in that instance the system is free of causal influences, whether from its own past or from the rest of the universe. This, however, is inconsistent with the fact that systems far from equilibrium are in constant interaction with their environment, and that their dynamics—as Prigogine well realizes—is critically determined by this interaction.

The assumption of a bifurcation process guided purely by the random play of fluctuations in the system (i.e., by an entirely unconstrained selection of the post-bifurcation dynamic regime) is inconsistent also with the empirical finding that, though the selection process in individual systems and instances seems random, it nevertheless manifests statistical regularities in large ensembles of systems and processes. Unless this were the case, the arrow of time in nature would not have a definite direction. Prigogine's principal endeavor is to demonstrate that time's arrow is a fundamental feature of the natural universe; he is well aware of the statistical bias in large ensembles of bifurcating systems. Indeed, his efforts to demonstrate that time—statistical irreversibility—is present already at the quantum level, led him to the concept of large-scale Poincaré systems—systems with a continuous spectrum and resonances—that are in constant interaction with their milieu.

(The application of these nonintegrable systems in quantum physics requires that the breaking of time-symmetry should occur already at the elementary level, rather than being, as John von Neumann suggested, a consequence of the act of measurement.[4]) Yet Prigogine did not sofar relinquish the idea of unconstrained freedom in the unstable system in regard to the selection of its post-bifurcation regime.[5] This is both contradictory in theory and inconsistent with observation. As just noted, while the selection of a new dynamic regime in a bifurcating system is individually undetermined, in large ensembles it is (stochastically) macro-determined. Rather than non-causal selection, there is a subtle but statistically significant constraint in the selection process by the system's environment, that is, by 'the rest of the universe'. In a nonintegrable Poincaré system this constraint can be real even if it is weak enough to be significant only in large populations. The phenomenon that appears as freedom in the individual case is in fact a statistical constraint manifested in the ensemble.

Independently of the choice of theoretical concepts, it should be clear that constant and universal interaction among phenomena cannot be suspended without fracturing the evolutionary process into divergent trajectories that would ultimately yield incoherence among its products. Coherence in the overall process requires constant universal interconnection among its elements. Given such interconnection, the sufficient repetition of a basic set of algorithms can produce complexity characterized by order and coherence. This is demonstrated in computer simulations. When an array of elements interacts with one another in a limited number of ways, the entire interacting system tends to fall into a limited number of states—for example, typically 100 for 10,000 interacting genes, in Stuart Kauffman's calculations.[6] The argument, derived from computer simulations, is compelling. In regard to its application to nature it presupposes interconnections throughout the order-generating system. With such interconnection, the conceptual requirements of an evolutionary process capable of generating divergence as well as convergence are met.

The relevance of fields

Most scientists are ready to admit that, in principle, 'everything is connected with everything else' but, beyond quantum field theories, where 'the rest of the universe' enters into the calculation of the state of a given elementary particle, the connections seldom inform the calculation of empirical events. Specifying and computing the interconnections that obtain among systems on various levels of complexity and evolution remains a challenge to natural science as a whole. In the view maintained here, the challenge can be met; the conceptual foundations for doing so will be outlined below. We begin here with the 'foundations of the foundations', that is, the indicated concept of interconnection.

If an event A at one point in space is connected with an event B at a different point, we must assume that A and B are interconnected by a continuous (causally or functionally correlated) matrix that has physical reality much like A and B. A physically real connecting matrix of this kind is best conceptualized as a field.

The concept of field has a considerable history: the need to link events at different points in space was known already to the Greeks. In modern times the corresponding need arose out of Newton's theory of gravitation. Gravitational effects make for a link between cause and effect, yet if the mysterious postulate of 'action at a distance' was to be avoided, some medium was required to transmit the causal effect. Beginning in the eighteenth century, physicists began to interpret gravitational action as action in a gravitational field. This field was assumed to be built by all the existing mass-points in space and to act on each mass-point at its specific spatial location. It was this notion that in 1849 Michael Faraday used to replace direct action among electric charges and currents with electric and magnetic fields produced by all charges and currents existing at a given time. In 1864 James Clerk Maxwell stated the electromagnetic theory of light in terms of the field in which electromagnetic waves propagate at finite velocity, and in 1934 Einstein pointed to Maxwell's concept of field as the most profound and fruitful transformation in our concept of reality since Newton.[7]

Today, quantum physicists view particle interaction in terms of quantum field theories. Quantum fields are more than theoretical abstractions: none of the known elementary particles could be seen as part of the furnishings of the physical universe except as manifestations of underlying energy or probability fields. Electrons, for example, are mathematically defined as point particles (particles without spatial dimension), but such particles act in space (in quantum field theory the interaction of electrons is described by an exchange of photons within the electromagnetic field) and they can do so only if they are in some sense *in* space. Hence, logically, electrons can be viewed as point-events in spatially extended fields. (This, of course, creates a conceptual problem that is not resolved to this day: a physically real field cannot have a dimensionless point except as an arbitrary singularity. Einstein himself was dissatisfied with this concept: although in 1927 he published a paper with Grommer in which he demonstrated that quanta can be treated as mathematical singularities in the electromagnetic field, he disliked the arbitrariness of the concept and did not publish further theoretical investigations of quanta.)

The actuality of quantum fields is underscored by the fact that the photons exchanged in electromagnetic field interactions are virtual particles without independent existence apart from their interactions within the fields. The same goes for the exchange particles created in the interaction of quarks. As electrons interact by the exchange of photons, so quarks interact by the exchange of gluons. In the case of the gluon force (also called 'color force'), the effects do not diminish with distance. On the contrary, quantum chromodynamics predicts that the force—and hence the number of gluons—increases proportionately to the distance between the interacting quarks. This would be a complete anomaly unless the space between the quarks was 'filled' by an extended dynamic field.

There are also other physical phenomena that presuppose the presence of underlying fields: they include the cloud-like charges of electrons (composed of quarks and antiquarks as well as of gluons exchanged between the quarks), and the transformation of quarks as a result of interactions (in weak interactions the quark changes 'flavor' but not 'color,' while in strong interactions, when a quark absorbs or emits a gluon, the quark changes 'color' but not 'flavor'). Theories in contemporary physics, including established theories such as general relativity's spacetime, as well as innovative and as yet controversial constructs such as string and super-string theory, are pure mathematical abstractions without the assumption of continuous fields underlying and interconnecting the phenomena.

Temporal field-connections

Fields do not only link events in space; linkages in time may be due to field-phenomena as well.

In classical science time-connections were ascribed to an unbroken chain of causes and effects. Universal laws of motion combined with rigorous causal transmissions enabled scientists to trace effects to prior causes with mathematical precision. For example, when a ball rolls down on an inclined plane, the speed and acceleration of the ball is determined by the law of gravitation plus the size and weight of the ball (assuming negligible friction between ball and plane). The law of gravitation was held to be a constant, entering into all things the same way. But the size and the weight of the ball were known to be variable: they served to specify the initial conditions of the process. The combination of constant gravitation with variable initial conditions made possible the precise description and accurate prediction of the motion of the ball along the plane.

Temporal connections through dependence on initial conditions extended logically to the very beginning of time: the initial conditions of every process could be seen as the effect of prior causes that, in turn, were the effect of still prior causes. Consequently classical science envisaged an unbroken causal chain stretching back to the hypothetical first instant when the universe was set in motion. Given that the laws of motion themselves were held to be space- and time-invariant, the initial conditions that reigned at that hypothetical instant were considered to have predetermined everything that took place thereafter. It enabled Laplace to declare that to a demon of unlimited computing ability in full knowledge of the current state of the universe, all past and future states would be evident.

This grand vision is no longer affirmed in science. By the first decade of this century the determinism of classical mechanics was discarded, and time linkages through chains of processes with initial-condition dependence had to be discarded. A probabilistic universe cannot be 'caused' by its past; at the most, specific events could leave traceable impressions on a limited range of subsequent events.

However, contemporary science knows other than deterministic forms of temporal linkages between phenomena. The relevant concept is a field, endowed with memory. Memory is not necessarily an anthropomorphic concept. Although in humans it is associated with mind and consciousness, memory can exist independently of these factors. The simplest of living organisms conserves some impressions of its environment: it has some form of memory although it does not possess a nervous system capable of mind and consciousness. Even an exposed film has memory: it 'remembers' the pattern of light of various intensities that reaches its surface through the camera lens; and the computer that processes the text now being written also has memory—and even a form of logic and intelligence—although it is not likely to have a conscious mind.

While there are several types of memory in nature, the one that appears to have greatest potential in connecting phenomena on the required universal scale is that which is associated with the hologram. The holographic principle has been known since 1946 when Dennis Gabor discovered it in his search for a more efficient microscope.[8] As used by scientists and engineers, holography is an artificial process, created for specific purposes. The process rests, however, on a physical principle that is likely to occur in nature. If it occurred in association with a field, it would endow that field with memory.

In a hologram information is recorded in a distributed fashion.* Because all parts of the holographic plate receive information from all parts of the photographed object, the full 3-D image can be retrieved by reconstructing the wave interference patterns stored on any part of the plate—although the smaller the part used in reconstructing the information the fuzzier the resulting image. This means that, since two or more parts of the holographic plate can be viewed simultaneously, observers on different locations can retrieve the same information simultaneously.

In addition to being distributed, holographic information storage is extremely dense. A small portion of a holographic plate can conserve a staggering variety of wave interference patterns. According to some estimates, the entire contents of the US Library of Congress could be stored on a holographic medium the size of a cube of sugar.

The here noted properties of holographic information storage and retrieval suggest that a field capable of interconnecting phenomena in nature is likely to function in the holographic mode. Interconnections in space call for the simultaneous availability of information at different spatial locations; and the distributed nature of holographic information storage responds to this requirement. Interconnections in time, if occurring on a universal scale, require in turn the conservation of a staggering amount of information; the holographic process satisfies this requirement as well. A universal field with holographic properties would constitute a medium with distributed read-out potential and quasi-unlimited information-storage capacities. *Prima facie*, it is a good candidate for ensuring space- and time-connections in nature.

The evolutionary functions of a holographic field

A holographically information-encoding and -transmitting field would make events in space and time into a non-Markovian chain.** Though this would necessitate a radical

---

*A hologram consists of a wave-interference pattern produced by two intersecting beams of light stored on a photographic plate or film. One beam reaches the plate directly, while the other is scattered off the object to be reproduced. The two beams interact, and the interference patterns encode the characteristics of the surface from which one of the beams was reflected. As the interference pattern is spread across the entire plate, all parts of it receive information regarding the light reflective surface of the object.

**In a Markovian-chain the elements $x_1, x_2, ... x_n$ are defined by mutually dependent random variables in such a way that predictions about the next link in the chain (x_{n+1}) can be made uniquely on the basis of a knowledge of the last link (x_n); in a non-Markovian chain, on the other hand, such prediction requires a knowledge of all links $x_1, ... x_n$.

departure in the computation of spatiotemporal events (a dynamic analysis of the internal structure of spacetime would have to be undertaken, for example, in terms of the 'infinitely small neighborhoods' of mathematical non-standard analysis), a non-Markovian chain of events would have significant order-generating properties. The past evolution of events in the chain would in a literal sense '*in-form*' their further evolution. When compared with a blindly groping trial-and-error process, this would make for a vast saving of evolutionary time.

The time-saving due to non-Markovian interconnections can be illustrated with an example cited by Fred Hoyle.[9] Suppose, said Hoyle, that a blind man is trying to order the scrambled faces of a Rubik cube. He is handicapped by not knowing whether any twist he gives the cube brings him closer to or further from his goal. In Hoyle's calculation his chances of achieving a simultaneous color matching of the six faces of the cube are in the range of 1: 1 to 1: $5 \cdot 10^{18}$. Consequently if the blind man works at the rate of one move per second, he will need $5 \cdot 10^{18}$ seconds to work through all possibilities. This length of time is not only more than his life expectancy: it is more than the age of the universe.

The situation changes radically if the blind man receives prompting during his efforts. If he receives a correct 'yes' or 'no' prompt at each move, the laws of probability show that he can unscramble the cube at an average of 120 moves. Working at the rate of one move per second, he will thus need two minutes, and not up to 126 billion years, to reach his goal.

The above example illustrates the difference that a certain kind of interconnection—in this case the feedback of information relevant to the prior state of the system—makes in an otherwise random process. If the connecting feedback is both perfect and compelling, the reduction of the number of decision-points needed to reach a goal can be dramatic. If it is neither perfect nor compelling, the time-reduction will not be dramatic, but could still be significant. Even an occasional and non-compelling 'prompt' could speed up random trial-and-error developmental processes. It could, for example, make the evolution of organic species fall within acceptable time-frames. (Even a partially random evolutionary process, we shall argue in Chapter 4, would considerably exceed the roughly 4 billion year timespan that was available in the biosphere.)

We would not expect that an outside intelligence 'prompts' evolution in nature, suggesting viable moves and averting dead-ends. That would be overshooting the mark: evolution would then be extremely rapid and fail-safe. It is neither. However, systems in space and time are still likely to be prompted by some variety of interconnection, since if they were at the mercy of chance, the known time-frames would be vastly exceeded. And the connecting prompt need not come from a supernatural agency: it could be generated in the system itself. The understanding the process that could generate it may hold the key to understanding the emergence of order in nature.

A caveat has to be entered here. Evolution in the universe, unlike the unscrambling of a Rubik cube—or any fixed-goal process—is open-ended: in each step it opens up more alternatives than it closes. It must, therefore, involve more than a self-generated prompt connecting the system with its own past states. If it were only that, the hand of the past would weigh ever heavier in the iterations of evolutionary processes; genuine innovations would be progressively filtered out. That is why, as we shall show in Chapter 2, the universe cannot be a 'system of habits' as in Sheldrake's theory. This problem does not beset a holographically generated 'prompt,' for holographic information can be in the form of superposed wave-patterns in multiple dimensions. Multidimensional signals can fit not only each system to its own evolutionary past, but entire hierarchies of systems to each other. Systems organized as nested hierarchies can thus move toward increasing order through the fine-tuned organization of their parts within the various levels of their multi-level structure.

The fifth field

Given the existence of a holographic information field, we could expect nature to evolve toward consistently ordered diversity without exceeding the known timeframes. Before examining this proposition in more detail, we should pose the fundamental question: does a holographic information field actually exist in nature?

Contemporary physics knows four varieties of universal field: these are the gravitational, the electromagnetic, and the strong and the weak nuclear fields. (According to grand-unified theories all four fields originated as a single 'super-grand-unified force' in the very early universe—the currently observed fields separated out subsequently, by spontaneous symmetry-breaking.) But it is not likely that the known fields would account for the kind of spacetime connections we require. The nuclear fields are local forces of interaction; they do not interconnect phenomena across wide stretches of space and time. Gravitation and electromagnetism are both universal fields, but it is not clear how they could provide the kind of subtle preferential prompts that are required if we are to account for the timely emergence of diversified yet ordered systems. There is, moreover, evidence of interconnections among phenomena that transcend the known limits of space and time (e.g., quantum particle nonlocality and environmentally adapted mutations of a massively 'systemic' kind, among others). Gravitational and electromagnetic fields of the known kind cannot be responsible for such connections. We should entertain the possibility that a 'fifth universal field' may exist in nature.

Physical roots of the fifth field

The physical roots of an information field that would provide universal interconnections among phenomena are not known with any certainty. Reasoned hypotheses can, however, be advanced and explored. Here we shall follow the lead of scientists such as David Bohm, who have looked in an analogous context to the quantum vacuum, the potential-energy field that defines the ground state of the universe. The possibility that the vacuum furnishes the indicated 'fifth field' merits a closer look.

The first thing to note is that the quantum vacuum exhibits so-called zero-point energies (ZPEs) — energies that persist even at temperatures close to absolute zero, when all other forces vanish. Hence the vacuum is not empty space, but a fluctuating field filled with zero-point energies. It is the successor to the nineteenth century concept of the luminiferous ether.

In its time, the ether concept made eminent sense: it explained how objects can influence each other beyond direct physical contact. The idea of an invisible medium that would fill space and convey effects over distance was proposed already by Descartes, who used it to explain the propagation of light and heat. Subsequently the ether was held to transmit not only light and heat, but also gravitational, electric, and magnetic forces. Solid objects were assumed to move through it, and in so doing to produce some level of friction. A.F. Fresnel produced detailed and experimentally testable calculations of the 'ether-drag,' and in 1881 Albert Michelson began a series of experiments to test his drag-coefficient. But the series of ingenious experiments, concluded in 1887 with E.W. Morley, showed no ether-drag whatever.

At first the physics community was reluctant to surrender the ether concept and sought alternative explanations: some spoke of a 'conspiracy of natural law' that would prevent the observation of motion relative to the ether. Then Einstein's theory of relativity permitted the calculation of physical effects without taking into account an ether drag: the computations referred to changes in the relative position of points in spacetime rather than to the motion of single points. The place of a universal reference frame, filled with a mechanistic medium, was taken by relativistic spacetime, described in geometrical terms.

In the course of this century, physicists replaced the notion of an ether-filled plenum with that of a cosmic vacuum. They reasoned that the ground state of the universe, though filled with energies associated with particles in their lowest energy state, is free of matter and gravitation: it approximates a vacuum. The bothersome ZPEs, it was thought, could be eliminated by the mathematical stratagem of 'renormalizing' the equations. This gave results generally consistent with the observed values. As Einstein's formulas did not require reference to an ether-like universal rest frame either, space in the absence of matter came to be seen as the quantum *vacuum*.

The assumption that ZPEs can be ignored in the calculation of physical effects may be unfounded, however. The vacuum's ZPF (zero-point field) may be more than a conveniently negligible curiosity of nature: it may convey various physical effects. This notion occurred to Michelson himself. In a paper published in 1881, he pointed out that the experiments did not call into question "the existence of a medium called the ether, whose vibrations produce the phenomenon of heat and light, and which is supposed to fill all space." The fact that the interpretation of the ether produced by Fresnel was disproved, Michelson noted, should not be taken as proof that there is no medium that fills space and time and transmits a variety of effects—gravitational, electromagnetic, and possibly still others.[10]

Though in Einstein's special theory of relativity the ether is replaced by a reference frame that is relative to the observer, Einstein himself did not disregard the implications of an enlarged concept of the ether. In 1924 he said, "In a consequent and coherent field theory, elementary particles constitute particular state spaces ... In this way all the objects are included again in the ether concept."[11] At around the same time he speculated that there may be a guiding field (*Führungsfeld*)—or, as he ironically phrased it in a letter to Bohr, a *Gespensterfeld* (ghost field) associated with the motion of photons.[12]

Einstein's intuitions are borne out by recent work at the innovative edge of the physics community. Some physicists, among them Manfred Requardt of the University of Göttingen and Ignazio Licata of the University of Sicily, now view the vacuum as a physical expression of spacetime. They consider quantum mechanics as a 'coarse-grained' theory of the more fundamental vacuum level. In Licata's view this level is 'reticular spacetime,' and it functions as an ultra-referential structure in which absolute deformations are described by the stochastic metric tensor and express deviations from isotropy and homogeneity in the Lorentz-invariant background. Thus Lorentz transformations are actual physical effects created by the spatiotemporal motion of matter.[12] Thomas Bearden, in turn, considers the electrostatic scalar potential as an n-dimensional stress in the quantum vacuum, where n is equal to, or greater than, four. In Bearden's theory the vacuum is equal to energy-filled spacetime: it is a highly charged cosmic medium. The virtual state of this medium determines all that emerges into physical reality as vectorial and matter-bound energy.[14]

Independently of speculative theories at the frontiers of physics, contemporary work in cosmology suggests that the vacuum is a significant energy field. It is known to be the originating source of matter in the universe: when sufficient energy is injected into the state of the negative energy particles predicted in Dirac's equation, the particles shift from a virtual state in this 'Dirac sea' into a real state in spacetime. It is also known that the vacuum is the ultimate sink of the matter particles that were synthesized in the universe. In Stephen Hawking's theory at the 'event horizon' of black holes one particle of a pair of particles escapes into surrounding space, while its antiparticle twin is sucked into the black hole, where it decays into the vacuum.

However, the quantum vacuum is more than the source and the sink of matter; recent evidence indicates that it can also be an active influence on the behavior of matter particles. For example, the zero-point energies of the vacuum interact with electrons around nuclei with effects that become observable when the electrons transit from one energy state to another: the photons they emit exhibit the 'Lamb-shift'—a frequency slightly shifted from its normal value. There is also the Casimir-effect, produced as the vacuum's zero-point energies create a radiation pressure on two closely spaced metal plates. Between the plates some wavelengths of the vacuum field are excluded, thus reducing the energy density with respect to the field outside. This creates the pressure that pushes the plates inward and together.

Innovative work in physics discloses ever more interactions between matter and the quantum vacuum. In the mid-1970s, Paul Davies and William Unruh argued that constant-speed motion through the vacuum would exhibit its spectrum as isotropic, whereas accelerated motion would produce a thermal radiation that breaks open the directional symmetry. The Davies-Unruh effect, too small to be measured with physical instruments, prompted scientists to investigate whether accelerated motion through the vacuum would produce further effects.

A recent investigation focused on the force of inertia. Originally defined as the property of a material object to either remain at rest or to move uniformly in the absence of external forces, inertia became part of the classical laws of motion (Newton's second law, $F = ma$). In this guise it became a fundamental quantitative property of matter, without, however, disclosing how it is associated with material objects. Ernst Mach suggested that inertia should be related to all matter in the universe, and Einstein attempted to integrate this principle into general relativity, but neither of them could offer a convincing demonstration. Recently Haisch, Rueda and Puthoff advanced a mathematical theory that exhibits inertia as a Lorentz force originating at the subelementary level and producing opposition to the acceleration of macroscopic objects.[15] The accelerated motion of physical objects through the vacuum produces a magnetic field, and the particles that constitute the objects are deflected by this field in accordance with the Lorentz formulas. The larger the object the more particles it contains, hence the stronger the deflection and greater the inertia. Inertia, in this theory, is a form of electromagnetic resistance arising in accelerated frames from the spectral distortion of the vacuum zero-point field. It follows that, if the inertial and the gravitational mass are identical and indistinguishable (Einstein's principle of equivalence), mass itself is generated in the process of interaction between accelerated charges and the zero-point electromagnetic field. Thus the vacuum's ZPF may be the truly fundamental element of the physical universe, the generating source not only of microparticles (under the aspect of the Dirac sea), but also the seemingly intrinsic properties of 'matter': mass, inertia, and gravitation.

According to a related and as yet likewise controversial finding, the vacuum also interacts with the photons propagating in it. Such interaction was rejected when the Michelson-Morley experiments showed no trace of an ether-drag on light propagation; but it may be that this decision was premature—the negative results may have been due to a misinterpretation of the results. Already in 1913, Georges Sagnac came up with different findings: the speed of light, he showed, does not remain invariant in a rotating frame of reference but varies with the direction of rotation. He maintained that these results support the classical theories of light of Huygens and Fresnel and prove the existence of an ether.[16] Sagnac's interpretation was subsequently contested, among others by Paul Langevin. But Langevin's interpretation was questioned in turn by Herbert Ives. Ives applied Poincaré's principle of relativity to the results of the Michelson-Morley experiments and came up with the 'rod-contraction-clock-retardation ether theory.' By elaborating the Lorentz equations of motion within a universal reference frame, the theory accounts for the experimental results usually cited in support of special relativity without the assumption of relativistic spacetime.[17]

More decisive evidence in regard to photon propagation was presented by Ernest Silvertooth. In 1987, on the one-hundredth anniversary of the Michelson-Morley experiments, he published experimental results that demonstrate that the wavelength of light varies with the direction of its propagation. Unlike Sagnac's experiments, which indicate that special relativity's light-velocity constant does not apply to rotating frames of reference, Silvertooth's experiment shows that the constant also fails to apply to light travelling in a straight line. The Earth, it appears, moves in space with an absolute velocity. Silvertooth's value for this velocity (378±19 km/sec) matches the independent astronomical determination of the Earth's motion relative to the cosmic background radiation (365±18 km/sec), as well as Monstein's anisotropy of cosmic ray distribution.[18] It may thus be that what the Michelson-Morley experiments have shown is that the average of the back-and-forth velocity of light within a given reference frame is constant, and this is as special relativity requires. They did *not* show, on the other hand, that the one-way velocity of light would be likewise constant irrespective of the motion of the observer.*

The origin of ZPEs in the vacuum is another unresolved question. Either these vacuum energies were fixed arbitrarily at the birth of the universe as part of its boundary-conditions, or they are being ongoingly generated by the motion of charged particles. Harold Puthoff undertook to demonstrate the latter. Puthoff calculated the properties of the radiation from particles by quantum fluctuations in spacetime (based on the electromagnetic radiation emitted by charged particles; this is known to drop off in space as the inverse square of the distance from the source). As the average volume distribution of charged particles in spherical shells surrounding any given point source is proportional to the area of the shell—as given by the square of the distance—the sum of radiations from surrounding shells yields a high-energy density radiation field. Puthoff identified it as the

---

*This variation, negligible at ordinary velocities, becomes important at high speeds. In a spaceship travelling through space at 95 percent of the speed of light, a photon moving in the same direction as the ship (from the back to front) would move 40 times slower than a photon travelling in the opposite direction. (Paul LaViolette, personal communication.)

vacuum zero-point field. His calculations show that the absorption and reemission of ZPF radiation by a ZDF-driven dipole oscillator yields a local equilibrium process: the radiation field generated by the ZPF-driven dipole just replaces the radiation absorbed from the ZPF background, doing so on a detailed balance basis with regard to both frequency and angular distribution. The feedback loop of charged particles generating the field, and the absorption of that field by the particles, is self-regenerating on the large scale: the local zero-point energy background experienced by a given charge is due to radiation from the motion of charged particles in the zero-point field throughout the rest of the universe. According to Puthoff's model, the energies of the zero-point field throughout the universe are continuously generated by the motion of quanta, and the sum of the motion of all particles in the universe in turn 'drives' the motion of quanta, producing a 'self-generating cosmological feedback cycle'.[19]

Last but not least, we should note that, independently of difficult and as yet unresolved questions concerning the origins of zero-point energies, the constancy of the speed of light, and the 'matter-like' properties of mass, gravitation, and inertia, the vacuum proves to be a significant factor in the equations that describe the quantized field obtained in the unification of electromagnetism with quantum theory. As work underway since the 1960s in stochastic electrodynamics (SED) demonstrates, when fluctuations in the vacuum are accepted as *a priori* givens, many of the puzzling facets of quantum behavior can be resolved with classical calculations. Haisch *et al.* point out that, though it is premature to claim that all quantum phenomena could be explained by SED, it is possible that such a claim would be justified one day.[20] One could then accept the laws of classical physics as providing a basically correct description of physical reality, provided the ZPF is included as a basic element of that reality.

Though in the space of this review we could not take account of the full breadth and depth of current vacuum-related theorizing, we have presented evidence to the effect that the interaction of the vacuum with the observable furnishings of the universe is a significant physical phenomenon. We thus have good reason to explore whether the vacuum may be the physical basis of the universal field that ensures the universal interconnections we have said to be required for a coherent process of physical and biological evolution.

References

1. Ilya Prigogine, *Thermodynamics of Irreversible Processes*, Wiley-Interscience, New York 1967 (3rd ed.).

2. Ilya Prigogine and Isabelle Stengers, *Order out of Chaos: Man's new dialogue with nature*, Bantam Books, New York 1984, p.16.

3. Ilya Prigogine and Isabelle Stengers, *op.cit.*, pp.169-70.

4. Ilya Prigogine, 'Why irreversibility? The formulation of classical and quantum mechanics for nonintegrable systems' *International Journal of Quantum Chemistry* (1994);
Mario Costagnino, Edgard Gunzig, Pascal Nardone, Ilya Prigogine and Shuichi Tasaki, 'Quantum cosmology and large Poincaré systems (mimeo), 1994.

5. Personal communication, November 1994.

6. Stuart Kauffman, *The Origins of Order: Self-Organization and Selection in Evolution.* Oxford University Press, Oxford 1993.

7. Albert Einstein, *The World As I See It*. Covici-Firede, New York 1934.

8. Dennis Gabor, 'A New Microscopic Principle,' *Nature*, **161** (1946).

9. Fred Hoyle, *The Intelligent Universe*, Michael Joseph, London, 1983.

10. A.A. Michelson, 'The Relative Motion of the Earth and the Luminiferous Ether,' *American Journal of Science*, **22** (1881).

11. Albert Einstein, *Proc. of Schweiz. Naturforschungs Gesellschaft*, **105** (1924).

12. Niels Bohr, in *Albert Einstein, Philosopher-Scientist*, P. Schilpp (ed.), Cambridge University Press, London 1970, 205-206.

13. Ignazio Licata, 'Dinamica Reticolare dello Spazio-Tempo' [Reticular dynamics of spacetime], Inediti no.27, Soc. Ed. Andromeda, Bologna 1989;
Manfred Requardt, 'From "Matter-Energy" to "Irreducible Information Processing" — Arguments for a Paradigm Shift in Fundamental Physics,' *Evolution of Information Processing Systems*, K. Haefner (ed.), Springer Verlag, New York and Berlin 1992.

14. Thomas E. Bearden, *Toward a New Electromagnetics*, Tesla Book Co. Chula Vista, CA, 1983.

15. Bernhard Haisch, Alfonso Rueda, and H.E. Puthoff, 'Inertia as a zero-point-field Lorentz force,' *Physical Review A*, **49**.2 (February 1994).

16. George Sagnac, 'The luminiferous ether demonstrated by the effect of the relative motion of the ether in an interferometer in uniform rotation,' *Comptes Rendus de l'Académie des Sciences*, Paris, **157** (1913).

17. Herbert Ives, 'Light signals sent around a closed path,' *Journal of the Optical Society of America*, **28** (1938);
    'Revisions of the Lorentz transformations,' *Proceedings of the American Philosophical Society*, V, **95** (1951);
    —, 'Lorentz-type transformations as derived from performable rod and clock operations,' *Journal of the Optical Society of America*, **39** (1949),
    —, 'Extrapolation from the Michelson-Morley experiment,' *Journal of theOptical Society of America*, **40** (1950).

18. Ernest W. Silvertooth, 'Experimental detection of the ether,' *Speculations in Science and Technology*, **10** (1987);
    —, 'Motion through the ether,' *Electronics and Wireless World*, May 1989;
    —, 'A new Michelson-Morley experiment,' *Physics Essays*, Vol. 5 (1992).

19. Harold A. Puthoff, 'Source of vacuum electromagnetic zero-point energy,' *Physical Review* A, **40**.9 (1989).

20. Bernhard Haisch, Alfonso Rueda and Harold E. Puthoff, 'Beyond $E = mc^2$' *The Sciences*, November/December 1994.

Chapter Two

BASIC CONCEPTS OF QUANTUM/VACUUM INTERACTION (QVI)

> *We are seeking for the simplest possible scheme of thought that will bind together the observed facts.*
>
> Albert Einstein, *The World As I See It*

The task before us is to advance the simplest possible scheme of thought capable of uniting the *prima facie* disparate domains of physical and of living nature. Such a scheme calls for an interactive evolutionary dynamics that is neither fully deterministic, nor punctuated by fully random events. The required dynamics is probabilistic but oriented; its probabilities constrained by order-generating interconnections among the evolving systems. Such interconnections, we have noted, are best conceptualized in terms of continuous fields, and the required properties of the fields are best satisfied by the holographic mode of information storage and transmission. The physical roots of the indicated holofield, in turn, are most reasonably sought in the Dirac sea of the quantum vacuum.

The specific objective of this Chapter is to sketch out those postulates of the integrative scheme that would describe the relevant properties of the quantum vacuum. These are the properties in virtue of which the vacuum could interconnect phenomena so as to produce a self-referentially randomness-mitigating evolutionary process.

This enterprise is necessarily speculative on at least two counts: first, because the required properties of the vacuum have first to be 'invented'; and second, because they concern an element in physical reality that is intrinsically unobservable. It may seem that we have in mind a free invention regarding an inaccessible part of reality—and this, sceptics may point out, amounts to engaging in metaphysics.

The above claim would be too harsh, however. It would apply with equal force to a very large part of contemporary physics, both in the Planck-domain of the quantum, and in the cosmic domain of cosmology and cosmogony (e.g., to the quantum probability state, black hole evaporation, strings and superstrings, grand- and super-grand-unified forces, etc.). Of course, the fact that some theories in contemporary science are speculative does not give license to engage in similar speculation in other domains without further warrant. There are cases, on the other hand, where sufficient warrant *is* given. One such case is the one we confront here: the need to describe the properties in virtue of which the quantum vacuum could function as a universal randomness-mitigating interconnecting field. The alternatives to creating a hypothesis in this regard are to find another explanation for the nonlinear yet stubborn emergence of order in nature—or to ascribe the entire matter to final causes and intentional designs. We do not see a satisfactory solution for the former option, and we shall not take the latter—not for being unreasonable, or even for being necessarily untrue, but for being beyond the bounds of natural science.

Neither unobservability nor the free invention of basic premises are legitimate obstacles to scientific theory-construction. Unobservability is not equal to unverifiability: in-themselves unobservable phenomena can have observable effects, in which case verification can proceed from observed effects to unobserved—but rigorously constructed—causes. Apart from this manner of proceeding, science would be confined to the domain of direct observability; a domain that, even when the power of human sense organs is extended by the use of telescopes, microscopes, and other probes, is presently only a small fraction of the domains of nature investigated by scientists. Invention, in turn, is not an unacceptable, or even an exceptional, method in the natural sciences: the basic premises of scientific theories, as Einstein said, are always the fruits of disciplined imagination. They are not works of fiction by that token, for they are methodically postulated and then rigorously tested, in terms of their deduced consequences. Theory construction, though imaginative, must yet respond to criteria such as simplicity, internal consistency, coherence in regard to 'neighboring' theories, and empirical testability. The latter is indeed the hallmark of scientific (as contrasted with metaphysical) speculation—valid hypotheses must be open to confirmation as well as to disconfirmation. Falsifiability, Popper's specification of the testability criterion, is generally regarded as the crucial hallmark of scientific theories.

The 'simplest possible scheme of thought that will bind together the observed facts' cannot be limited to the directly observed or even the instrumentally observable facts: if it were, it would be neither simple nor schematic. As our reflections in Chapter 1 have shown, the simplest possible scheme capable of binding together the relevant facts involves a specific bias in stochastic system-development processes; and that bias in turn requires reference to a holographic information field that underlies and embeds the observed phenomena. While such a scheme is necessarily speculative in itself, it can nevertheless be confronted with empirical evidence. This is a task we shall undertake in the Chapters that follow.

Here we outline the two sets of postulates that make up the basic scheme. One set concerns the properties of the field to which holographically pattern-conserving and transmitting properties are ascribed; and the other shows the application of the postulates to micro- and macro-scale systems in the domains of observation and experiment.

The vacuum postulates

We begin the construction of the basic scheme by ascribing the properties to the quantum vacuum by virtue of which it would function as a universal interconnecting field.

Constructing a scheme that would exhibit the vacuum as a universal interconnecting field requires attributing a substructure to it. This is consistent with the evidence already in hand. The relevant finding is the high degree of complexity that underlies the interactions of quanta. This suggests either that quanta themselves are compound entities, with an internal structure that accounts for the specific complexity of their interactions, or that the structure of the field in which they are embedded has the required degree of complexity. Both these assumptions have been explored by theoretical physicists. Bohm, for example, assumed that the structure of particles—somewhere between 10^{-16} m and 10^{-35} m—may be complex enough to respond to information that he took to originate with the pilot wave (the quantum potential Q). There is, however, no independent evidence for assuming that quanta would be complex entities in themselves. It is known on the other hand that the quantum vacuum is a cosmically extended electromagnetic field. In this field the interaction of massless charges creates 'matter' in the form of mass, which in turn creates gravitation and produces the attractive and repulsive forces associated with nuclear fields. This field does have a substructure, though mainstream physics views it as homogeneous and isotropic, filled with purely random fluctuations (*Zitterbewegungen*).

The assumption that the massless charges that constitute the fundamental units of the observable universe would be embedded in a vacuum field with some type of substructure is intrinsically reasonable. The question is mainly whether that substructure interacts in significant ways with the phenomena that appear in it. In stochastic electrodynamics (SED) the assumption that it does so interact produces remarkably accurate accounts of quantum phenomena, without requiring the complex auxiliary assumptions of contemporary quantum mechanics.

In view of considerations such as these, we choose the second option: quanta, in this view, are embedded in a complex field. That field could, in principle, be non-randomly structured, that is, it could be information-rich. Let us consider this question.

Quanta as solitons

The basic and perhaps most remarkable property of the quantum vacuum is its energy density. The predicted radiation spectrum of the zero-point field (ZPF) rises proportionally to the cube of the frequency of its radiation. Wheeler calculated that, assuming that quantum laws hold all the way to the Planck-length of 10^{-35} m, the vacuum's energy-content is equivalent to 10^{94} g/cm^3. This magnitude, according to Bohm, exceeds all energy bound in matter by a factor of 10^{40}. If this energy were associated with mass, the resulting gravitational potential would reduce the curvature of the universe to an order of magnitude several dimensions smaller than the nucleus of an atom. However, the ZPF may not gravitate. Quantum geometrodynamics shows that at the Planck-length of 10^{-35} m the oscillations of this super-energetic gas break up the structure of the spacetime continuum, giving rise to sporadically connecting and disconnecting distinct spacetime segments; and the resulting quantum foam—'superspace' in Wheeler's terminology—is a pure massless charge-flux.

When fluxes in the vacuum cross the energy-threshold of particle-creation ($6 \cdot 10^{-27}$ erg/sec) its virtual particles transform into 'real' particles. Real particles in space and time seem to enjoy an independent existence: they are endowed with corpuscular, in addition to wave, properties. Nevertheless, it now appears that mass is not a fundamental property of quanta, but a product of its interaction with the ZPF.[1] 'In-themselves', quanta are best viewed as massless charges propagating in a continuous electromagnetic field—non-Markovian singularities in a complex manifold. In this perspective quanta approximate solitons (solitary waves), rather than constituting discrete spatiotemporal entities.

Solitons, though they give the appearance of discrete objects, are part of the medium in which they occur. The phenomenon was first reported to the British Association for the Advancement of Science in 1845, when J. Scott Russell re-counted riding beside a narrow channel of water and observing a wave rolling with great speed, "assuming the form of a large solitary elevation, a rounded, smooth and well defined heap of water, which continued its course along the channel apparently without change of form or diminution of speed."[2] Similar phenomena have since been observed in a number of cases involving turbulent and nonlinear media. Solitons appear in impulses of the nervous system, in complex electrical circuits, in tidal bores, in atmospheric pressure waves, in heat conduction in solids, as well as in superfluidity and superconductivity. The Great Red Eye of the planet Jupiter, though seemingly a detached object, is a soliton produced by Jupiter's turbulent surface. Solitons move along defined trajectories, and if their trajectories meet, they may deflect each other. It is thus with good reason that a number of physicists now view them as a dynamic metaphor for the behavior of quanta.

As the initial step in the construction of our unified scheme, we define quanta as soliton-like waves within the quantum vacuum. This 'embeddedness' of quanta is obscured if we mathematically renormalize the quasi-infinite energy values of the vacuum: the description we get then is that of a wave packet moving in empty space. If, however, the virtual-particle gas of the vacuum is properly taken into account, the description will be that of quanta propagating as soliton-like waves in a fluctuating and possibly information-rich energy field.

The question concerns the information-richness of the quantum vacuum, more exactly, of its zero-point electromagnetic field. As noted in Chapter 1, Puthoff, as most other physicists, views the ZPF as homogeneous, isotropic, and Lorentz-invariant. This is to satisfy the finding that constant motion through the ZPF does not give rise to asymmetries (though accelerated motion produces the distortions predicted by Davies and Unruh and viewed as the physical basis of inertia by Puthoff, as well as Rueda and Haisch). The query that needs to be posed, however, is whether asymmetries of motion detected by physical instruments are the only conceivable kinds of effects that the field would produce on observable phenomena. The concept of the ether was discarded for failing to produce effects of this kind, yet this should not have done away with the concept of a continuous underlying field that would function as a charge-conveying and information-transmitting medium. As already remarked, there is observational evidence for the existence of such a field, and the possibility that the quantum vacuum would function as that field cannot be discarded solely on the consideration that constant-speed motion through the ZPF does not produce physically measurable effects. The effects could perhaps be produced, but not identified as effects *of* that field.

Secondary waves in the vacuum

In this study we embrace the concept of an information-rich subquantum field as the most reasonable heuristic device in regard to universal interconnections in nature. Of the two sets of postulates that spell out the hypothesis, the first set concerns the creation of secondary force-fields within the ZPF. As is well known, classical electrodynamics predicts that a fluctuating electric charge emits an electromagnetic radiation field. In their interaction with the ZPF, electrically charged quanta are believed to produce secondary electromagnetic fields, and these field must be universally extended. (The thus produced fields create a weak force of attraction among the quanta, equivalent to—and in the view put forward by Puthoff, Rueda and Haisch, actually identical with—gravitation) The energy associated with the fields gives rise to fluctuations in charged particles that propagate relativistically, at or near the speed of light. *Zitterbewegung* is assumed to be random, satisfying the tenet of homogeneity, isotropy and Lorentz-invariance for the ZPF. However, our hypothesis is that the secondary fields created by the motion of quanta have a nonisotropic, nonhomogeneous and non-Lorentz-invariant component. This component consists not of the familiar transverse electromagnetic waves, but of longitudinally propagating scalar ('Tesla') waves. In this regard the motion of charged particles through the ZPF approximates the action of a monopole antenna: it alternately charges and discharges local regions of the primary electromagnetic field. The thus triggered longitudinal waves alternately compress and rarefy the virtual-particle gas of the vacuum. In consequence the structure of the ZPF becomes mediated by secondary scalar fields, generated by quanta. We envisage the ZPF as a structured field with a scalar-mediated electromagnetic spectrum. The consequences of this postulate can be spelled out.

In regard to its scalar component, the ZPF becomes a continuum of which each point is defined by a corresponding magnitude. At each point of the field the nonrandom flux is a local scalar wave within the massless charge spectrum. The magnitude at a given point is an n-dimensional virtual-state flux. The field is a continuum of stresses and potentials. Its stress energies can be expressed in terms of geometrodynamics as the electrostatic scalar potential $\emptyset(phi)$, where the quantity $\emptyset_o$ measures the amount of work that must be performed against a unit charged mass in order to push it in from infinity against the charged field potential.

We now consider the specifics of the two kinds of secondary fields within the vacuum. The first thing to note is that the secondary scalar waves are not of the kind that satisfy D'Alembert's equation, that is, they are not similar to light and sound waves. This is because the characteristic feature of a D'Alembert equation is the occurrence of a second time-derivative term of the wave amplitude, and generally such a term is a consequence of the inertial properties of matter. In a medium such as the quantum vacuum, these properties would not apply; vacuum waves are better represented by fundamental equations that contain only first-order time-derivative terms. But there is only one kind of first-order time-derivative equations governing linear wave propagations and these are Schrödinger wave equations. We postulate that the scalars mediating the ZPF approximate Schrödinger waves.

In regard to at least four factors, this additional postulate satisfies the memory-functions we require to be associated with an information-rich ZPF. First, Schrödinger waves are linear, thus allowing interfering wave-trains to superpose, conserving rather than destroying phase information. Second, unlike holographic wave patterns composed of D'Alembert waves that can only be recorded on plates, interference patterns created by Schrödinger-type waves require the entire medium in which the waves propagate. This allows for vastly more information storage by Schrödinger-wave holograms than by holograms based on D'Alembert waves. Third, the diffraction of Schrödinger waves through a fixed hologram can create time-varying information, whereas only time-invariant information can be recorded with D'Alembert type interference patterns. And fourth, items of time-varying information injected into the holographic medium through a system of independent pointlike sources can be recovered in a Schrödinger-type hologram in the proximity of the sources, whereas information recovery in D'Alembert-type holograms is possible only by focalization devices for the diffracted waves.

We next consider the velocity of Schrödinger-type secondary scalar wave-propagations in the vacuum. Schrödinger waves generated by periodic emissions are known to propagate with velocities proportional to the square root of their specific frequencies. In the vacuum these velocities cannot be held limited by the constant that applies to the propagation of charged masses in the electromagnetic spectrum. In relativity theory the light constant c— presently estimated at 299,748±15 km/sec—functions as an axiom of invariance: in a normal coordinate system the speed of light is to be chosen so as to remain constant regardless of the curvature of spacetime; and the unit of time is to be chosen so that the speed of light within the local system of coordinates becomes equal to one. However, if Silvertooth is right and c varies with the motion of the observer relative to the light source (cf. Chapter 1), the space and time invariances of special relativity no longer hold.*
Instead, we obtain a physical basis for the observed value of c: the finite electromagnetic permittivity/permeability of the vacuum. If the value of c is inversely proportional to the square-root of the product of the vacuum's electrostatic permittivity and magnetic permeability ($c = 1/\sqrt{\varepsilon_0 \mu_0}$), assuming that the electric and magnetic components refer to comparable vacuum constraints, we find that $\mu_0 = 1/\sqrt{c}$ and $\varepsilon_0 = 1/\sqrt{c}$. This gives a physical basis for the limited velocity of charged masses but leaves massless charge propagations unaffected.

Scalars do not propagate as classical electromagnetic waves. In a seminal, though largely neglected, paper published in 1903, E.T. Whittaker has shown that longitudinal waves propagate with a finite velocity that is not the same as that of light: it may be enormously greater.[4] Scalars are longitudinal waves, and it is now recognized that their propagation is proportional to the medium's mass-density. Mass-density defines the local electrostatic scalar potential of the vacuum. It is a variable quantity, higher in regions of dense mass, in or near stars and planets, and lower in deep space—a variation due to the increase in vacuum flux intensity by the accumulation of charged masses. Consequently scalars travel faster through matter-dense regions of the vacuum than in deep space, much as longitudinally propagating sound waves travel faster in a dense medium such as water than in a thin medium such as air (and as Whittaker's longitudinal gravitational waves travel faster than light waves).

---

* Einstein himself did not ascribe unlimited validity to the postulate of the constancy of light: in 1911, and again in 1916, he provided a detailed demonstration that in a gravitational field the speed of light is a function of location. The special theory, he admitted, has validity only insofar as one can disregard the effects of gravitational fields on the phenomena.[3]

The interaction postulates

Vacuum–quantum interaction

For the scalar-mediated ZPF to produce an effect on observable phenomena, it must interact with the soliton-like quanta embedded in the vacuum. We assess this interaction as internal to the vacuum as an information-rich universal energy field. Within that field, the interaction of scalars and quanta approximates a two-way Fourier transformation: the vacuum encodes the coefficients of the interfering scalar wavefronts produced by the motion of quanta. In so doing the vacuum carries out the equivalent of the forward Fourier-transform: it translates a pattern from the spatiotemporal to the spectral domain. In the inverse transform—from the spectral to the spatiotemporal domain—the interference patterns encoded in the vacuum 'in-form' the motion of quanta in space and time. With some simplification we can say that there is an ongoing *read-in* by quanta of their spatio-temporal motion into the vacuum, and a similarly ongoing *read-out* of the corresponding information.

The process can be illustrated with the translation that occurs between the motion of vessels and the surface of the sea. When a vessel creates waves on the sea's surface, it creates Fourier-transforms of its impact on the waters. The waves are not merely passive representations of the motion of the vessels that triggered them: they are structuring factors of the surface that are an active influence on the motion of other vessels. However, in the vacuum the wake created by a 'vessel' is an active influence not only on the motion of other vessels, but on its own motion as well. This is because, unlike in the ordinary sea, the waves triggered by quanta propagate faster than the quanta that created them. As a result each quantum is affected by the secondary scalar waves it has itself triggered within its system of coordinates.

The two-way translation between quanta and the quantum vacuum is universal in scope but specific in occurrence: quanta retranslate from the vacuum only those transforms that match their own quantum states. Such specificity is due to the fact that in Fourier transformations the reverse transforms are the exact inverse of the forward transforms. Hence a high specificity of retranslation is typical of holographic processes. When laser beams of different frequency (color)—are used to record a specific interference pattern on a holographic plate, the wavefronts superpose in a multidimensional interference pattern.

When the plate is illuminated with a laser beam of a specific color, only that image appears that has been recorded by a beam of that color.

Retranslation selectivity ensures that quanta are not overwhelmed by the information conveyed through the scalar-mediated ZPF. Quantal states are affected only by the spectral transform of their own wave functions even if, due to the high velocity of scalar wave propagations, in matter-dense regions the feedback affects all quanta simultaneously. In the here introduced simplified terminology we can say that each quantum 'reads out' from the vacuum only the information that it has itself 'read in'.

Micro- and macro-scale applications

The above postulates need to be applied to two seemingly disparate types of systems: microscale systems, of the order of Planck's constant; and macroscale systems, consisting of large ensembles of microscale components.*

As just noted, the motion of quanta generates scalar patterns in the ZPF, and the thereby modified topology of that field interacts with the motion of quanta in the corresponding quantum state. This tenet, as we shall see, can shed light on various puzzling quantum phenomena, including nonlocality and nondynamic correlation.

First, however, we should note the temporal dimension of the interaction of micro- and macro-scale systems with the scalar-mediated ZPF. Our considerations on this score are purely theoretical: they rest on the assumption that the secondary scalars created in the ZPF are not subject to attenuation by the action of any known force or process. Though the information carried in the waves is produced by the motion of quanta, given linear Schrödinger-type waves, further quantal motion will only produce a retranscription, and not a cancellation, of the preexisting information. (In a multi-dimensional superposition of linear wave-interference patterns the information contained in each component wave is integrated, and not destroyed.) This means that the information encoded in the substructure of the quantum vacuum has no presently known—or even readily conceivable—temporal limitation.

The hypothesis of constant interaction with enduring vacuum-based information selects a particular concept of the quantum. Quanta cannot be seen as subsisting in a 'pure' state,

---

* In this Chapter only basic concepts are discussed; their exploration in regard to empirical phenomena is the topic of Part Two.

independent of exchanges with the rest of the universe. Instead, they must be viewed as nonintegrable Poincaré systems with continuous spectra and constant resonance-based interactions. Poincaré resonances, as Prigogine has shown, limit the validity of trajectory description in classical mechanics and of wave function description in quantum theory: they introduce 'diffusive terms' breaking time-symmetry for a well-defined class of dynamical processes. In simple (individual) situations these terms can be neglected, and then the traditional formulations of classical and of quantum physics can be recovered. However, when a large number of systems is involved, these terms become significant: they introduce both probability and irreversibility into the description.[5]*

These considerations have important implications for our understanding of the quantum state. The universe as a whole is logically viewed as a large-scale nonintegrable Poincaré system,[7] and if so, all systems within the universe must be seen as engaged in persistent interaction. The isolation of a quantum becomes an abstraction, even if in individual cases assuming it is a valid approximation. Fundamentally quanta are nonisolable constantly interacting entities in which constraints on degrees of freedom, introduced by interactions with the rest of the universe, though individually nonmanifest, appear on the level of ensembles. This tenet safeguards the formalisms of quantum mechanics in regard to individual systems, but qualifies them with interaction-generated probabilistic irreversibilities for populations of systems. For the latter, constant interaction with the rest of the universe cannot be neglected. Interaction between quanta and the energy field in which quanta are embedded must be affirmed, even if it is not apparent in individual instances.

We can now move beyond the Planck-domain to macroscale systems. Prima facie it appears that in macroscale systems vacuum (and other) fluctuations are damped by large-scale regularities that obey dynamical laws. Interaction effects could appear nevertheless, as Poincaré resonances manifested in populations of the component quanta are amplified by the chaos-dynamics of the system. As generally known chaos (more exactly, states of deterministic, strange-attractor-regulated chaos) is highly initial-condition dependent. Un-

---

* The reason for this is that in large-scale Poincaré systems the spectral representation is valid only on the level of statistical ensembles. As a result irreversibility, as Boltzmann already noted, is a 'population property' that does not apply to individual systems. That in individual cases isolation provides a valid approximation so that classical description in terms of trajectories, and quantum mechanical description in terms of wave functions, applies does not obviate the fact that in ensembles of systems Poincaré resonances limit degrees of freedom and introduce additional constraints, leading to small deviations from the classical trajectories amplified by diagonal singularities. Hence in regard to ensembles description must go from trajectories to probability distributions, and from wave functions to density matrices.

measurably small changes in initial conditions—or, in persistent systems, in parametric conditions—are registered in chaotic systems, as strange attractors amplify the changes into dynamic inputs to the evolutionary trajectory. We suggest that chaotic-state macroscale Poincaré systems are often sufficiently interactive to produce measurable interaction-effects. These interactions involve not the quantum state of the individual component, but the $3n$-dimensional configuration-space of the whole multiquantal system. As a result, in dynamically indeterminate ('chaotic') states macroscale systems can be 'in-formed' with the ZPF-conveyed Fourier-transform of their $3n$-dimensional configuration-space.

The above tenet applies with particular force to living systems. These are systems that persist far from thermodynamic equilibrium, in states that are inherently unstable and ultrasensitive. This condition has been theoretically recognized since mid-century, but it is only in the last two decades that observations have highlighted its functional consequences. In their far-from-equilibrium condition, living systems prove to function on a knife's edge between dynamical stability and chaotic instability. Their states require mapping by a variety of attractors, including chaotic ones. Even under relatively stable conditions, when turbulent states are not manifest, chaotic attractors may be present in a latent form, somewhat like recessive genes. Periodically entire organisms, as well as particular organs and organ systems, enter prevalently chaotic states. Even the heart, a paradigm of stable regularity, 'dances' when it is healthy. Though it normally remains within 60-80 beats per second, the beats vary greatly from second to second and minute to minute, and the complexity of its variations cannot be readily predicted. On the other hand, before a heart attack the EKG, as Pool noted, is typically "an epitome of stable regularity, with a mostly flat line interrupted every second or so by a quick up-and-down blip marking the beat."[8] Also the neural nets in the neocortex are in a predominantly chaotic state; chaotic behavior serves as the essential ground state for the neural perceptual apparatus. Investigators in neural nonlinear dynamics note that in many physiological areas 'the more chaos the better'—even though it is often difficult to say why. Nonlinear processes occur at critical phases in the coupling of signals across cell membranes, enabling the organism to respond to minute, quasi-unmeasurably fine alterations in its internal and external environment.

The living organism is highly sensitive to ambient radiation fields and amplifies the inputted effects through the dynamics of chaos. Sensitivity extends to energies and radiations well below the thermal threshold of chemical reactions (expressed as kT, a function of the Boltzmann constant and absolute temperature). There is evidence of cellular and biomolecular sensitivity to low frequency electromagnetic radiation (so-called ELF waves), as well as to electric gradients below thermal noise levels. Responsiveness in humans and monkeys to electric gradients in tissue fluid occurs already in the range of 10^{-7} to 10^{-8} v/cm, considerably below the 10^5 v/cm electric barrier of the cell membrane potential. In molecules or segments of molecules resonant vibrational or rotational interactions occur with electromagnetic waves in the range of 10 to 1,000 GHz. The growth of yeast cells, in turn, is finely tuned to applied field frequencies around 42 GHz, with successive peaks and troughs at intervals of about MGz. The sharpness of the tuning even increases as field intensity decreases: responses occur with incident fields as weak as 5 picowatts/cm. Investigators find that the imposed EM fields can be active even at intensities near zero.[9] These experimental findings indicate that the condition of the organism is not dependent uniquely on the equilibrium thermodynamics of thermal energy exchanges and tissue heating. Its interactions involve quantum states and resonant responses.

The sensitivity of the living state is apparent both in regard to level of excitation and the time-scale of the excited state. For example, in the human eye visual perception is initiated by a protein absorbing a photon. When that occurs, the protein is excited into a high-energy metastable state that creates the series of chemical reactions that gives rise to visual perception. The photon-absorption process itself lasts only a few picoseconds, and the various dynamical stages the follow occur on femtosecond timescales—even though the eye itself can only respond within a few hundredths of a second.

As a consequence of recent findings the investigation of biomolecular processes is moving beyond biochemistry, into quantum physics. The physics involved *inter alia* in biophoton research, bioenergetics and bioelectromagnetics, is a major emerging factor in the investigation of life phenomena.

Related hypotheses

The above concept of quantum/vacuum interaction (henceforth abbreviated QVI), was put forward as the 'simplest possible scheme' capable of unifying the observed facts in the physical and in the life sciences. It can, and should be, assessed in relation to analogous endeavors by contemporary theoretical scientists. To begin this comparative assessment we review here pertinent schemes advanced by David Bohm, Henry Stapp, and Rupert Sheldrake.

Bohm's implicate order

Much like QVI, Bohm's theory of the implicate order refers anomalous empirical phenomena to an interaction between quanta and the quantum vacuum; more precisely, to a factor that originates in the quantum vacuum and affects the motion of quanta. The quantum vacuum, in the guise of the implicate order, is the generative ground of all physical (and even of psychical) events and phenomena.[10]

According to Bohm, the totality of the manifest world derives from the implicate order as an explicate sub-totality of stable recurrent forms. Because all things are given together in the implicate order, there are no longer any chance events in nature; everything that happens in the explicate order is the expression of order in the order of the implicate realm. The implicate order acts on the explicate order by determining the motion of quanta. The key factor is the quantum potential. The same as the gravitational constant G, the quantum potential Q pervades spacetime. Q originates in a subquantum domain beyond space and time. Out of this enfolded holographic order it emerges as a 'pilot wave' that guides the motion of quanta. The effect, for Bohm, depends uniquely on the form, and not on the energy of the holographic waveforms. Hence, unlike the forces of gravitation and electromagnetism, Q does not diminish in space or attenuate in time.

These concepts are generally in agreement with the scheme outlined here. However, notwithstanding this overlap, there are areas of fundamental. The first of these concerns the nature and origins of the order-generating medium. For Bohm, everything that emerges into the explicit order is already given in the more fundamental subquantum dimension. The implicate order itself does not evolve: it merely subsists: it is an essentially Platonic realm of eternally given Forms. This is not the case in regard to the vacuum field postulated here: the interfering wavefronts that pattern this holofield are a direct result of the behavior of quanta in space and time. Matter and subquantum field, two basic aspects of physical reality (we would not name them separate dimensions), interact, and co-evolve as a result of their interaction.*

A more profound difference concerns the philosophy of physics implied by the theory of the implicate order. Bohm assumes that in addition to the wave function of quantum

---

* In his last years Bohm espoused a concept that brought his theory closer to the QVI concept. Prompted by criticism among others by Sheldrake, he envisaged the quantum potential associated with the totality of particles in space and time as generating the 'superimplicate order' which, through the implicate order, feeds back to create the evolving quantum potential that in-forms each particle. (David Peat, personal communication.)

theory, there are further factors that pertain to the physical world as it is classically conceived. The universe is governed by deterministic laws applying to particles and to fields. Every particle is accompanied by a wave that satisfies the Schrödinger wave function: this 'pilot wave' determines the quantum potential, which in turn determines the particle state. The structure of particles—possibly between 10^{-16} m and the Planck-length of 10^{-35} m— is said to be complex and subtle enough to respond to information from the pilot wave. Hence particles do not intrinsically possess both particle-like and wave-like properties: the observed wave-like properties follow from the general effect of the quantum wave-field on their structure. Thanks to this postulate, Bohm can explain probabilistic quantum events in reference to deterministic processes. In his 'ontological interpretation' the wave function is not merely an expression of potentiality for possible observations, but exists in reality.

Except for the above noted determinism (which is reminiscent of hidden variable theory)—and an unnecessary insistence on the separate reality dimension of the holofield, Bohm's concepts are fundamentally compatible with those of QVI. In our concept as well, the in-formation of the systems that make up the sphere of manifest reality occurs through a constant exchange with the universe's holographically interconnecting field. This exchange does not occur by means of a deterministic pilot wave, however, but through the continuous spectra of large-scale Poincaré systems.

Stapp's 'Heisenberg quantum universe'

Stapp's relevant work is a transdisciplinary theory based on the extension of the range of applicability of quantum mechanics to macroscopic events.[11] This concept, Stapp notes, dispenses with Bohm's classical variables while retaining the idea that the probability distribution that occurs in quantum theory exists in nature, and not just in the mind of the observer.

A realistic interpretation of the detection event was suggested by Heisenberg himself. "If we want to describe what happens in an atomic event" he wrote in a collection of papers published in 1985, "we have to realize that the word 'happens'... applies to the physical not the psychical act of observation, and we may say that the transition from the 'possible' to the 'actual' takes place as soon as the interaction between the object and the measuring device, and thereby with the rest of the world, has come into play; it is not connected with the act of registration of the result in the mind of the observer."[12] (This contrasts with an earlier view, according to which the objective reality of particles has 'evaporated' into the mathematics by which particle behavior is represented.[13])

In the realistic interpretation adopted by Heisenberg, the wave-like aspects of nature are represented by probabilistic states in conjunction with the associated Heisenberg operators. The basic operators are spacetime points, and their 'expectation values' in any Heisenberg state give definite values defined for all spacetime points from the infinite past to the infinite future. All wave-like properties of nature are embedded in these values. The wave-like aspects are interpreted as 'objective tendencies' for the occurrence of 'actual events'. The latter correspond to the particle-like aspects of nature; they are the counterparts within nature of the events that are customarily viewed as the 'collapse of the wave function'. As every actual event represents an objectified collapse of the wave function, it is not a subjective occurrences in the mind of the observer but a sudden change in the objective Heisenberg state of the universe.

When this conception is applied to all spheres of nature, within as well as beyond the Planck domain, we get the Heisenberg quantum universe complete with large-scale nonclassical effects. In this universe the quantum probability distribution, together with its undetermined resolution, makes for a complete representation of reality. The representation discloses that the evolution of the physical world proceeds by an alternation between two phases: a gradual evolution via deterministic laws that are analogous to the laws of classical physics; and the periodic occurrence of sudden quantum jumps. The latter actualize one or the other of the various macroscopic possibilities generated by the deterministic laws of motion. The jump occurs as interactions decompose the quantum probability distribution into well separated branches. The detection event actualizes one of the alternatives and eliminates the others. The actualized alternative may also be a macroscopic event, distinguishable at the level of direct observation.

Stapp claims that, if the nonclassical mathematical regularities identified by quantum theory are accepted as characteristics of the world at large (a world he terms more 'idea-like' than 'matter-like'), then we "appear to have found in quantum theory the foundation for a science that may be able to deal successfully in a mathematically and logically coherent way with the full range of scientific thought, from atomic physics, to biology, to cosmology, including also the area that had been so mysterious within the framework of classical physics, namely the connection between processes in human brains and the stream of human conscious experience."[14]

This claim, while in line with the intended scope of the here presented unified scheme, leaves open two fundamental questions. The first is 'where to, in the quantum jump?' As Stapp admits, these jumps are not strictly controlled by any known law of nature. Contemporary quantum theory treats them as random variables in the sense that only the statistical weights of the jumps are specified: the actual choice of this or that event is not accounted for in the theory. The occurrence or nonoccurrence of the jumps, and the choices they effect remain entirely random with respect to the preexisting possibilities. However, when it comes to extending quantum mechanics to macroscopic phenomena such randomness is a flaw: if we leave the way the quantum state evolves to chance, we cannot give a coherent account of the progressive emergence of order in the macroscopic domains. The kind of orders we observe, as already noted in Chapter 1, cannot come about in a random universe; they exceed permissible timeframes even if random events intersperse otherwise nonrandom processes.

The second question is whether, on Heisenberg's premises, probabilistic states would actually come about in the macro-domain. This is not clear. When we conceptualize quanta in terms of Poincaré systems, it becomes questionable whether even a photon travelling from the source that emits it to the counter that registers would be free of interaction. It is far more doubtful that a macroscale systems such as an organism would be ever sufficiently isolated from its environment to be in a probabilistic state. In matter-dense regions of the universe detection events would be dense enough to prevent dynamical laws from creating the probabilistic alternatives among which the quantum jumps would decide.*

---

* A possibility advanced by Roger Penrose and Stuart Hameroff may save the Heisenberg quantum universe in regard to the operations of the brain and consciousness—Stapp's principal concern. If Penrose and Hameroff are right, underlying the transmission of signals between nerve cells there is a complex network of microtubules arranged in arrays. The protein molecule that makes up a microtubule has a kind of pocket along its length, and a single electron can slide back and forth along this pocket. This would determine the way the protein configures itself and therewith the configuration and function of the microtubule. Such tubules appear to be extraordinarily good conductors of physical pulses: a pulse introduced at one end can propagate unchanged throughout its length. Moreover disturbances in neighboring microtubules display significant coherence, so that a vibration in one tubule can propagate through an array of tubules. By allowing a pulse to travel along the microtubule without interacting with the molecules in its wall, an array of microtubules could insulate a vibrating pulse from the pattern of interaction that underlies the brain's information processing networks. If so, the pulse would be conserved in the probabilistic condition of the state vector, and able to explore simultaneously a number of possible states within the brain. Though neither Penrose nor Hameroff can say how long the state function would persist and how its persistence would enable the brain to choose one among its several possible deterministic states, the mere possibility of such a choice within the brain may provide a conceivable basis for the phenomena of consciousness and free will.

Although the conceptual difficulties of the Heisenberg quantum universe prompt a search for alternatives, it is important that the scope for the latter does not reduce to a universe of classical (or quasi-classical) determinism. Another and more likely alternative is a universe where, in the microdomain, interaction with the topology of the quantum vacuum in-forms the quantum state and where, in the macrodomain, the same interaction in-forms dynamically indeterminate states in macroscale systems. The resulting stochastic process can build ordered systems in space and time without either the constraint of full determinism, or the unrestraint of pure chance.

Sheldrake's formative causation

Further agreement and disagreement surfaces with the QVI scheme when we consider Rupert Sheldrake's concept of the morphogenetic field. The principal area of agreement is evident: in both schemes fields perform an effective formative role. For the life scientist Sheldrake this role is evident first and foremost in the generation and regeneration of organic form.[13]

The idea of morphogenetic fields dates back to the 1920s, when Alexander Gurwitsch used it to account for processes in embryology and developmental biology. Around the same time Paul Weiss applied biological fields to explain processes of regeneration in animals. Geometrical forms—explored among others by D'Arcy Thomson and Hermann Weyl—were connected with dynamical processes by Conrad Waddington and René Thom: they divided the biofield into geometrical zones of structural stability. More recently Brian Goodwin demonstrated the effect of biofields in growth processes in various organisms, most specifically in marine plants. Nevertheless, investigations of the biofield are somewhat exceptional in theoretical biology: for many researchers the field remains a heuristic device called for when there are no conventional explanations.

For Sheldrake, biological fields are more than heuristic devices: they have a reality of their own. In his view morphogenetic fields are continuously shaped and reinforced by previously existing organisms of the same kind. Hence living members of a species are linked with the forms of past members of the same species through morphic resonance, a process based on similarity of form or pattern. Morphic resonance is not limited to living organisms: it shapes formative processes in the realm of crystals, molecules, and atoms. In consequence Sheldrake's theory implies a universal principle of form and order in nature.

Though our scheme agrees that fields exist independently of the entities on which they act, it does not agree that the fields would convey their effects through morphic (or any other kind of) resonance. Matter/field interaction through resonance would reinforce the kind of form or order that has already appeared without allowing for significant deviations. Sheldrake's examples underscore this point: the more a given crystal has been synthesized the faster it will be synthesized in the future; the more a given behavioral routine has been learned by rats the faster other rats will learn the routine; the more organisms of a given species have existed, the more likely it is that organisms of that species will be generated again—and so on. On the one hand, this tenet offers an advantage: the forms and structures that emerge in nature are no longer at the mercy of chance as probabilities shift to favor outcomes that approximate the already existing forms and structures. On the other hand it also creates a difficulty. If the more a given structure or behavior has occurred in the past the more it recurs again in the future, genuine innovations could hardly establish themselves—they would be squashed by the heavy hand of the past before they could develop an effective morphogenetic field of their own.

Sheldrake is aware of the problem. If the universe is a system of habits, he asked, how do new patterns ever come into being—what is the basis of creativity? A theory of evolutionary habit demands a theory of evolutionary creativity. Could creativity on Earth be a product of the imagination of a Gaian mind? And could such an imagination, working throughout the natural world, be the basis of evolutionary creativity in nature the same as in the human realm?[16]

A further difficulty of Sheldrake's theory is the ad hoc character of the field that acts as the causal agent. It is not clear how a universal morphic resonance field would relate to what is already known about the physical universe. Resonance is a bona fide physical phenomenon (it is the reinforcement of the vibration of a body by the vibration of another body at or near its frequency), and in string and super string theory resonances have a major role. But there is no indication that this phenomenon would act independently of some form of energy. Yet Sheldrake suggests that there is a non-energetic form-creating field for every atom, molecule, crystal, or organism that has ever come into being. This means that there is a morphogenetic field not only for rats and rabbits, but also for quarks, and for the fermions and bosons made of quarks, and for the asteroids, stars, planets, galaxies and galactic clusters made of fermions and bosons. The entire cosmos would resonate with morphic fields, independently of the energies available for it. A similar claim is not required for the vacuum-interaction concept described in the QVI scheme: according to this concept the minute energies required to 'in-form' micro- and macroscale systems are either transferred from the scalar-mediated ZPF, or are available in the chaos-dynamics of the systems themselves.

Summary

The 'simplest possible scheme' presented in this Chapter responds to the requirement of providing a conceptual foundation for a transdisciplinary unified theory capable of integrating our understanding of events in the physical and in the living world. It suggests an explanation for the observed events in reference to matter/vacuum—more exactly, quantum/ZPF—interaction. The scalar-mediated and hence information-rich electromagnetic spectrum of the vacuum encodes the multi-dimensional spectral transforms of the state- and configuration-spaces of quanta and macroscopic systems of quanta. Interaction with the transforms reduces randomness in the evolutionary process: it injects the subtle yet effective bias that transforms the random collapse of the wave-function in microparticles, and the equally random bifurcations of macrosystems, into a self-consistent pattern of development. A chance-ridden change dynamic turns into an evolutionary process hallmarked by a measure of self-consistency, producing divergence as well as convergence in its trajectories and diversity as well as consistency in its products.

References

1. Bernhard Haisch, Alfonso Rueda and Harold E. Puthoff, 'Beyond $E = mc^2$' *The Sciences*, November/December 1994.

2. J. Scott Russell, *Report on Waves*, British Association for the Advancement of Science, 1845.

3. Albert Einstein, 'Einfluss der Schwerkraft auf die Ausbreitung des Lichtes', *Annalen der Physik*, **35** (1911), 898-908;
   —, *Über die spezielle und die allgemeine Relativitätstheorie*, Akademie Verlag, Berlin, 21. Auflage 1916, 61.

4. E.T. Whittaker, 'On the partial differential equations of mathematical physics', *Mathematische Annalen*, **57** (1903), 333-355.

5. Ilya Prigogine, 'Why irreversibility? The formulation of classical and quantum mechanics for nonintegrable systems,' *International Journal of Quantum Chemistry* (1994).

6. *Ibid.*

7. Mario Costagnino, Edgard Gunzig, Pascal Nardone, Ilya Prigogine and Shuichi Tasaki 'Quantum cosmology and large Poincaré systems (mimeo), 1994.

8. R. Pool, 'Is It Healthy to be Chaotic,' *Science*, **243** (1989) 604-607.

9. Del Giudice, E., G., S. Doglia, M. Milani, and G. Vitiello, in F. Guttmann and H. Keyzer (eds.), *Modern Bioelectrochemistry*. Plenum, New York 1986;

10. David Bohm, *Wholeness and the Implicate Order*, Routledge & Kegan Paul, London 1980;
David Bohm and B.J. Hiley, 'Non-relativistic particle systems.' *Physics Reports* **828** (1986).
—, *The Undivided Universe*, Routledge, London 1993.

11. Henry P. Stapp, *Matter, Mind, and Quantum Mechanics*, Springer Verlag, New York 1993.

12. Werner Heisenberg, *Physics and Philosophy*, Harper & Row, New York 1985, p.54.

13. Werner Heisenberg, in *Daedalus*, 87, 1958, 99-100.

14. Henry P. Stapp, 'Quantum Theory and the Place of Mind in Nature,' in *Niels Bohr and Contemporary Philosophy*, J. Faye and H.J. Folse, (eds.) (forthcoming).

15. Rupert Sheldrake, *A New Science of Life*, Blond & Briggs, London 1981;
—, *The Presence of the Past*, Times Books, New York 1988.

16. Terence McKenna, Rupert Sheldrake and Ralph Abraham, *Trialogues at the Edge of the West*. Bear & Co., Santa Fe, NM 1992.

PART TWO

EMPIRICAL EXPLORATIONS

Chapter Three

Q V I IN MICROPHYSICS AND COSMOLOGY

The objective of the QVI scheme presented in Chapter 2 is to provide a consistent and parsimonious explanation of the emergence of diversified yet ordered systems in the physical as well as the biological domains of observation and experiment. Reaching this objective calls for describing a factor of constant and universal interconnection in nature. This factor we have identified as the scalar-mediated zero-point field of the quantum vacuum. To test the heuristic power of the concept, we shall now insert the vacuum interaction-effect postulated in QVI into current theories in physics and biology. This is the program for the three Chapters that make up Part Two of this study.*

---

* Before embarking on this enterprise, we should remark that testing a hypothesis that refers to an in-itself unobservable domain of nature is necessarily indirect: it can only concern the inferred consequences of the hypothesis for the observable domains of nature. Such testing is feasible, but it does not yield a definitive true/false answer. As Einstein noted, the basic premises of scientific theories are products of the disciplined imagination and, as Mach, Duhem and Poincaré observed, an indefinite number of explanations (and hence theories) applies to any set of observations. These tenets apply with particular force to theories of in-themselves unobservable domains of nature. Thus, instead of categorical answers, we should look for probabilities of truth-value in reference to criteria such as heuristic power—in particular as regards otherwise anomalous observations—as well as simplicity, parsimony, and consistency with currently validated theories.

The QVI account of universal interconnection claims more than heuristic value for the transdisciplinary unification of physical and biological phenomena: it also claims validity within each of these domains independently of the other. Within physics the concept provides a reinterpretation of certain phenomena; and this reinterpretation could resolve otherwise anomalous aspects.

Over a century ago, James Clerk Maxwell wrote, "to fill all space with a new medium whenever any new phenomenon is to be explained is by no means philosophical, but if the study of two branches of science has independently suggested the idea of a medium, and if the properties which must be attributed to the medium in order to account for electromagnetic phenomena are of the same kind as we attribute to the luminiferous medium in order to account for the phenomena of light, the evidence of the physical existence of the medium will be considerably strengthened." Indeed, to fill all space with a medium such as the information-rich ZPF would not be 'philosophical'—unless research in two or more branches of science shows that the properties which must be attributed to that field in order to account for the observed phenomena with optimum consistency and parsimony are of one and the same kind, and are moreover consistent with what we must attribute to real particles in the zero-point energy state. The former of these possibilities will be elucidated as we continue to explore the basic scheme of quantum/vacuum interaction in various fields of science, while the latter is the topic to which we turn next.

Interaction effects in microphysics

For the greater part of this century, quantum theory has had difficulty in accounting for the phenomena it investigates with any measure of realism. The problems and anomalies have been accumulating for the past one hundred years. In the year 1900, Planck showed that energy radiates from a body in discontinuous packets called quanta; in 1905, Einstein proved that light, in addition to its known undulatory properties, possesses a corpuscular character. Then, in 1913, Bohr demonstrated that electrons move from one orbital trajectory around a nucleus to another without passing through intermediate stages; and in 1923, de Broglie postulated that quanta have irreducible corpuscular as well as wave properties. A few years later, in 1927, Heisenberg formulated the principle of uncertainty by which effective limits are set to our knowledge of an observer-independent quantum world, and Bohr was forced to the Copenhagen interpretation according to which the quantum is an 'elementary phenomenon' that cannot be known, and must not be speculated upon, beyond recording (and mathematically computing) its observations. The above term, John Wheeler later pointed out, is significant: it suggests that in speaking of quanta we are no longer speaking of objective, observer-independent reality. In Eugene Wigner's apt phrase, quantum theory had henceforth to do with *'observations'* — not with *'observables.'*

According to Wheeler, present quantum theory represents a state of complete knowledge about a dynamic system by a probability amplitude which is a complex number. The question that is not answered, as Wheeler himself points out, is a probability amplitude *of what?*[1] Bohr said that no elementary phenomenon is a real phenomenon until it is 'brought to a close' by an irreversible act of amplification, such as the clicking of a Geiger counter in the laboratory, or the blackening of a grain of photo-graphic emulsion. There are observations of these acts of amplification and assignments of the corresponding probability amplitudes with complex numbers, but no answers as to what lies behind the amplitudes. The wave function of the quantum state is not assigned to reality: it remains an expression of potentiality for various possible observations. As a result quantum physicists live in Alice's Wonderland where there are appearances of things — the grin of Cheshire Cats — but not the things of which they would be the appearances *of*.

Bohr and the Copenhagen school of quantum physicists ended by pronouncing an interdiction regarding speculation as to the nature of what the elementary quantum phenomenon may be 'in-itself.' The interdiction permitted work to progress in the mathematics of correlating the various observations without bothering with the 'philosophical' question as to what the observations would refer to (Bohr was reputed to have a sign over the door of his laboratory saying 'work in progress — philosophers keep out). Nevertheless, while many physicists abided by the Copenhagen interdiction, others did hazard explanations as to the nature of the reality that may underlie the 'elementary quantum phenomenon.'

The quantum-interpretation decision-tree

The choices that are available to post-Copenhagen theorists, while varied, are not infinite. They can be mapped in the form of a decision-tree.[2]

The decision-tree begins with a fundamental choice: should we inquire into the observer-independent reality of the quantum phenomenon, or ignore the entire ontological issue? If we opt to ignore it, we can proceed with experiments and observations without further speculative baggage. If, however, we decide to confront the issue, we find several alternatives at our disposal. The first alternative is to remain within the confines of quantum theory proper. In that case we may deny that there is a quantum world independent of the observer. If so, we must affirm that the observer has the eminent reality; the data coming to light in quantum experiments are folded into his experience. But we can also opt to strike out beyond radical quantum phenomenalism and allow for an observer-independent quantum reality. In that case we are logically obliged to clarify the nature of that world; in particular, why it is—or should appear to be— affected by our observations of it.

We may answer that the conscious mind of the observer acts on the quantum event: this is the position espoused among others by John von Neumann and Eugene Wigner.[3] We may also answer, however, that it is not the observer of the event, but the observed event itself, that decides the outcome: for example, the electron chooses its own state, as asserted by R.G. Jahn and B.J. Dunne.[4] A third alternative is that the observed universe divides into as many alternative universes as there are possible observed states; a fairly widespread view championed among others by H. Everett.[5]

If none of the above is satisfactory, we can retrace our steps along the decision-tree and take another branch. The alternative branch leads beyond the permissible horizons of the Copenhagen school along the lines maintained by Einstein: it views current quantum theory as essentially incomplete. In mid-century Bohm explored this branch in his famous, if intensely controversial, theory of hidden variables. Since then an entire array of field theories replaced the local determinism of hidden variable theory with the interactive determinism intrinsic to the ensemble of relations in which quanta are embedded. In S-matrix, bootstrap, and quantum field theories the state of quanta is referred to the ensemble of interactions that characterize the totality of the physical universe.

The question we confront at this point concerns the nature of this totality. Bohm suggested that beyond the sphere of observation and experiment there is another, non-temporal and non-spatial sphere where all that ever took place, and all that ever will, is given completely and permanently. As we have seen in Chapter 2, this interpretation is not fully satisfactory, but it is not the only alternative at this bifurcation of the decision-tree— we can also choose to postulate a uni-dimensional reality. The choice we confront then is whether generalize quantum laws so as to hold at all orders of magnitude, or whether to define the quantum state by the interaction of quanta with the quantum vacuum. The former alternative—Heisenberg's quantum universe—encounters the problem of randomness:

the choice of the deterministic state triggered by a 'decision-event' is not dynamically linked to the foregoing state of probability-superposition, so that the order-generating process lacks a vector toward convergence. The latter alternative is that which we shall explore here.

A search for overcoming the Copenhagen interdiction is more consistent with the generally realistic stance of natural science than abiding by the tenet that the quantum world is an elementary phenomenon. Einstein may have been right: quantum theory as presently formulated may be essentially incomplete. This view was reaffirmed by Dirac when he said that "present quantum mechanics is not in its final form. Further changes, about as drastic as the changes which we made in passing from Bohr's orbits to quantum theory proper, will be necessary. It is very likely that in the long run Einstein will turn out to be correct."[6]

An interpretation of quantum nonlocality and coherence

In the context of the QVI scheme, the particle's choice of a quantum state is not random: it is subtly but effectively influenced by the interaction of the particle with the vacuum. Certain phenomena of the quantum world—more particularly nonlocality and coherence—are non-anomalous in light of this concept.

We characterize the interaction between quanta and the quantum vacuum as a two-way Fourier transformation process. Every interference pattern in the vacuum feeds back to the corresponding spatiotemporal state- or configuration-space, obeying the principle that the inverse of the transform that translates from the spatiotemporal into the spectral domain translates from the spectral to the spatiotemporal domain. Consequently within a matter-dense region of spacetime, every photon, every electron, every atom, and every nucleon within every atom, is continually in-formed, if not with the wave-function of the entire region, then with that which corresponds to its own state- or configuration-space.

The above in-formation is not active as long as the quantum is in a non-interactive state. That condition, however, is an abstraction. In terms of that abstraction we may agree with quantum mechanics that the wave-function is properly described as a superposition of alternative possibilities of state. However, when the quantum is confronted with a decision event—and in the real world such confrontation is constant—its probabilistic vector state resolves into a deterministic state. The choice of that state is not specified in quantum mechanics: the latter can only give the weightings of the alternate possibilities. Constant quantum/vacuum interaction in continuous-spectrum Poincaré systems requires the choice to be consistent with the state- or configuration-space in which the system finds itself.

We can explore this concept in reference to the phenomena of nonlocality and coherence. The classic example is Young's double-slit experiment. Here a light is emitted from a source and is allowed to pass through a narrow slit in a screen. Another screen is placed behind the first, to register the rays that traverse the slit. Then, the same as if one allowed water to flow through a small hole, the light beam fans out and forms a diffraction pattern. The pattern shows the undulatory aspect of light, and is not anomalous in itself. But if a second slit is opened in the screen there is a superposition of two diffraction patterns, even if but a single photon was emitted. The waves propagating behind the slits form the characteristic interference pattern, with the wave fronts cancelling each other when their phase difference is 180 and reinforcing each other when they are in phase. It appears as if each photon would be passing through both holes at the same time.

Time and space seem to make little difference to the effect. In the 'cosmological' version of the experiment the interfering photons originate in a distant galaxy, emitted thousands of years apart. In one experiment the photons were emitted by the double quasi-stellar object known as 0957+516A,B. This is a distant quasar believed to be one object rather than two, its double image due to the deflection of its light by an intervening galaxy situated about one-fourth of the distance from the Earth. The deflection due to the gravitational lens action is large enough to bring together two light rays emitted by the quasar billions of years ago. Because of the additional distance travelled by the photons that are deflected by the galaxy, they have been on the way fifty thousand years longer than those that came by the direct route. But, although originating billions of years ago with an interval of fifty thousand years, the photons arriving in the laboratory interfered with each other the same as if they had been emitted merely seconds apart.

In order to avoid the conclusion that each photon is informed of the state of the other and chooses its path accordingly, quantum physicists normally assume that each photon, having both particulate and ondulatory properties, splits and takes both paths: it passes through both slits. It then recombines to produce the interference pattern on the screen. This view is consistent with the mathematics of quantum theory and has been held with few exceptions for over fifty years.

The standard view may need revision, however. An experiment reported by David Chalmers demonstrates that photons pass through only one of the slits, and then combine with photons passing through the other slit to produce the interference phenomenon.[7] In this experiment a laser light source is split into two beams and then recombined via half-silvered mirrors before reaching two slits in a screen. Interference in the form of alternating bands of light and dark is observed, as usual. Then a polarizing film is placed in front of the top slit, oriented so as to ensure that only vertically polarized photons can pass through that slit. A second polarizing film is placed in front of the bottom slit, oriented at an angle that allows only horizontally polarized photons to path through.

As we know, oppositely polarized beams cannot interfere with each other. If each photon, regardless of its polarization, passed through *both* slits, the opposite polarization would not prevent it from producing the interference pattern. But under conditions of opposite polarization the interference is not produced. The pattern is produced again when a third polarizing film is placed in the path of the photons emerging from the two slits, oriented at forty-five degrees to both, so as to remove the difference in their angle of polarization. With the angle of polarization harmonized, the photons interfere with one another.

Here a condition specified by Feynman has been satisfied for the reformulation of the laws of quantum mechanics. According to Feynman reformulation becomes necessary if the double-slit experiment is performed in such a way that the interference-pattern is obtained while the choice of slit through which the photon passes is determined. By introducing a specific polarization into the beam before reaching a given slit, individual-path polarizing experiments can determine through which slit a given photon can pass through. When they remove the incompatible polarization angles before the photons reach the registration screen, they again obtain the usual interference pattern.

It appears, then, that photons successively emitted by a given light source pass through only one of the slits and then interfere with each other. This suggests that not self-interference, but *memory* is at work. The previously emitted photons are no longer physically 'there', but their *traces* can be: the most reasonable interpretation is that the successively emitted photons interfere with the traces of the photons that were emitted previously. But how are these traces registered and conveyed? For a realistic interpretation a physical medium carrying the traces is required: a medium that transcends the limitations of relativistic spacetime. The scheme outlined in Chapter 2 furnishes the indicated medium in the guise of the scalar-mediated spectrum of the quantum vacuum.

We can now turn to the thought experiment proposed in 1935 by Einstein with Boris Podolski and Nathan Rosen (the 'EPR experiment'). Here a particle is split into two in identical quantum states and the halves are allowed to separate. A measurement is carried out on one (for example, of momentum) and a complementary measurement is to be carried out on the other (e.g., of position), thus showing that the Heisenberg uncertainty principle can be overcome. However, when in 1982 Alain Aspect succeeded in carrying out a physical testing of the thought experiment, it turned out that as a measurement is made on one of the particles, the superposed probability state is resolved (the 'wave function collapses') also in the second.[8] This, however, is not the point intended by Einstein; on the contrary, it places in question the law of maximum-velocity signal propagation in the theory of relativity.

Findings such as these prompted other thought experiments, including that which became known as 'Schrödinger's cat.' Erwin Schrödinger proposed that we take a cat and place it in a sealed container. We then set up a device which, entirely randomly, either does or does not emit a poisonous gas into the container. Thus when we open the container the cat is either dead or alive. Common sense would suggest that the cat died already when the gas was emitted—if indeed it was emitted—hence that it was either dead or alive before the container was opened. But this state of affairs is forbidden by quantum theory. As long as the container is sealed, there is a probabilistic superposition of states—the cat must be *both* alive *and* dead. It is only when the container is opened that the two probabilities collapse into one.

A similar thought experiment was proposed by Louis de Broglie. This time we take an electron rather than a cat, and place it in a sealed container. We divide the container, which is in Paris, and ship one part to Tokyo and the other to New York. Common sense would dictate that if we open the half-container in New York and find the electron, then the electron must have been in that half already when the container was shipped from Paris. But this state of affairs, like that which decides whether Schrödinger's cat is alive or dead, is forbidden by quantum theory. Each half-container must have a non-zero probability of harboring the electron. Then, the instant one of the halves is opened in New York— regardless of whether or not it contains the electron—the wave-packet defining the probability of the electron's existence is also reduced in Tokyo.

That spatially separated particles can be informed of each other's state constitutes the phenomenon of quantum nonlocality. Physicists have suggested various explanations of it. It is evident that the correlation is due either to a common system of coordinates embedding the particles, or to some other effect, such as the reversal of time (so that the observed particle could inform its as yet unseparated twin of the measurement that is being carried out on it).[9] The latter alternative has been given a twist by Huw Price, in the provision that it is the experimenter's intervention in the state of the particle that affects the prior state of that particle. This provides a realistic explanation of the phenomenon, but at the cost of assuming the existence of backward causation in the physical world.[10]

The alternative assumption—that of a system of coordinates that embeds both particles—requires that the coordinating system be specified and then identified in the physical universe. The QVI scheme proposes to do just that. Its coordinating system is constituted as an interactive process involving a two-way translation between particles and the quantum vacuum. Here the choice of the quantum state is not random—as in the standard quantum theoretical interpretation—but is linked with the rest of the universe through simultaneously interfering vacuum-based wave-transforms.

The scheme can also elucidate nondynamic correlation among electrons within atomic shells. In interpreting the physical events that underlie Pauli's exclusion principle (which states that no two electrons around a nucleus can be in a state of motion described by the same set of four quantum numbers), we can go beyond the oft-voiced—though wistful—remark that, here, too, each electron seems to 'know' in some way the quantum state of the others.[11] Instead, we can claim that in atomic shells electrons are interconnected by a signal-transmitting field. The wave functions of individual quanta within the configuration-space of the atom are quasi-instantaneously informed with the wave-function of the ensemble—an 'in-formation' that skews probabilities in the otherwise indeterministic quantum state toward that set of quantum numbers which corresponds to the position of each electron within the ensemble.

An analogous explanation accounts for the curious order that emerges in the electron shells of highly excited, so-called Rydberg atoms. The phenomenon can be best seen in the hydrogen atom. Under normal conditions, the sole electron of the hydrogen atom is tightly coupled with the proton in the nucleus. The behavior of the whole system is governed by the laws of quantum mechanics: the atom cannot enter any arbitrary energy state; it can only adopt discrete, that is quantized, energy levels. At low energies, the permissible levels are spread relatively far apart. But as the energy of the atom is increased, the electron moves farther from the proton and its permissible energy levels bunch closer together. At energies just below the critical threshold where the electron is stripped from the nucleus, the permissible bands would move close enough to form a continuum that can be described by the laws of classical mechanics.

However, Rydberg atoms exhibit a curious chaotic behavior. Mapping it requires that the phase space of the electron, normally represented in six dimensions (three dimensions for position and three for momentum), be reduced. Two dimensions can be readily eliminated, since the magnetic field introduced into the atom defines an axis of symmetry: the electron moves effectively in a two-dimensional plane where only the distances *along* the axis and *from* the axis are of relevance; motion around the axis is not. Out of this four-dimensional phase space one can take a three-dimensional slice, known as the energy shell: this permits the construction of a three-dimensional phase portrait of the electron's motion. The complexities of the three-dimensional phase portrait can then be further simplified by taking a so-called Poincaré-section (a fixed two-dimensional plane) through the energy shell. A two-dimensional representation is thus obtained, consisting of points at which the electron's trajectory intersects the plane's surface.

It is in this representation that the curious features of the Rydberg atom's behavior become evident. It turns out that, as other chaotic systems, the Rydberg atom cannot be decomposed to the motion of individual components: motion along one coordinate axis is coupled to motion along all the others. Moreover the distribution of the energy levels undergoes change. Whereas in the atom's basic state the energy levels are randomly distributed and are not correlated with each other, the energy levels of the Rydberg atom are nonrandomly distributed and strongly correlated. The continuum of levels toward which progressive excitation propels the atom breaks up: the energy bands become both highly separated and mutually linked. Yet there are no dynamical forces present in the atom that could account either for the new-found order exhibited by the energy band, or for their close correlation.[12] A meaningful explanation presupposes reference to constant (vacuum-based) interactions in large-scale Poincaré systems. In the magnetically induced chaotic state the effects of the interaction become manifest, since the atom is more sensitive than in its ground energy state. As a result its state-function exhibits the coherence we may expect contiguous solitons to exhibit within the vacuum's scalar-mediated zero-point field.

A closely related explanation can be essayed in regard to phenomena of coherence in superconductivity and superfluidity. The vanishing of electrical resistance in a superconductor—super-cooled helium and similar superfluids are known to penetrate without resistance through narrow capillaries and other channels—is attributed to the high degree of coherence among the electrons whose movement creates the current (the Schrödinger wave function assumes the same form for all the electrons). In supercooling the perturbation introduced at ordinary temperatures by the vibratory motion of the atoms in the lattices is removed, thereby restoring coherence. The current can thus pass through the conductor without resistance. This phenomenon can be likewise seen as an effect of the embeddedness of electrons within the information-rich vacuum spectrum. Given some degree of permeability/permissivity in that field (suggested independently by the finite speed of quantal propagations), charged particles in mutual proximity entrain each other. In the absence of thermal perturbation, they move coherently as soliton-like waves in a continuous underlying medium.

The Josephson effect (the correlation of superconductors at finite distances from each other) also points to a continuous underlying field; an interpretation that is less speculative than boson tunnelling across the intervening space. According to Del Giudice *et al.* the correlation effect obtains beyond the special case of superconductivity. Wherever there are neighboring systems of material entities—be they particles, atoms, or molecules—instant correlations may surface among them. A pair of nearby cells can act as a Josephson junction, and a set of identical cells can create an entire array of such junctions with the phase of the cells locked in. Since coherence within cells produces coherence among cellular assemblies, we have here a major factor ensuring the overall integration of the living system.[13] In this regard the interaction of micro- and macroscale systems with the holographic substructure of the vacuum spectrum emerges as a paramount process in both the physical and the biological worlds.

Sustained research may show the fruitfulness of referring the observed phenomena of the Planck-domain to the interaction of quanta with the spectral field of the quantum vacuum. In the context of this interaction the coherence of the quantum state is not an ad hoc given, and the state that results upon the collapse of the wave-function is not random in nature and unexplained in theory.

The quantum-vacuum interaction outlined here is innovative in specific detail, yet its philosophical foundations agree with a wide spectrum of opinion at the leading edge of physics research. We agree in particular with Bohm's thesis as regards the completion of quantum theory through consideration of subtle effects originating in the quantum vacuum, and with stochastic electrodynamics in integrating the fluctuations of the vacuum in the calculation of physical effects. The basic philosophy of the concept agrees with Wheeler's dictum: "vacuum physics lies at the core of everything."[14]

Interaction effects in cosmology

Anomalies persist in the macrodomain of cosmology as well; and here, too, the QVI scheme of constant universal interconnection may prove illuminating.

The problem of the constants

One of the fundamental problems of modern cosmology is that the magnitudes that make up the parameters of dynamical processes in the universe are improbably finely adjusted. Not only are the parameters that underlie the processes of life fine-tuned to the physical processes of the universe—as well they might be, since life emerged out of the physical background—also the parameters that set the prior physical processes are finely tuned to the subsequent processes of life. For example:
- The expansion rate of the very early universe was precise in all directions at a rate of better than one part in 10^{40}. Had it been less precise, cosmic background radiation would not be as uniform as it is. Yet it included small-scale departures from the large-scale uniformities - because, had these not been present, there could not be galaxies, stars, and planets in the wide reaches of cosmic space.
- The force of gravitation is precisely of such magnitude that stars can form and exist long enough to generate sufficient energy for life to evolve on suitable planets.
- The mass of the neutrinos, if not actually zero, is small enough to have prevented the universe from collapsing soon after the big bang due to excessive gravitational pull.
- The value of the strong nuclear force is precisely such that hydrogen can transmute into helium and then into carbon and all the other elements indispensable to life. If the value of that force were but two percent higher, all hydrogen in spacetime would have burned up before transmuting into heavier elements.
- The weak nuclear force has the exact value to allow atoms to be expelled in supernovae—and thus be available in next generation stars for building into the more complex elements that form the basis of life.
- The weak nuclear force also has precisely the value with respect to gravity that makes hydrogen rather than helium the dominant element in the cosmos. Had helium been

the dominant element, stars would not have existed long enough for life to have evolved on their planets and, for lack of hydrogen—the principal element of water—life as we know it could not have evolved.

As the above sampling indicates, the remarkable tuning of the universal constants concerns the values associated with the basic forces and fields of interaction, as well as with the mass and distribution of matter in cosmic space. These values, masses and distributions are precisely those that permit the evolution of increasingly complex matter-energy systems. This finding prompts a number of physicists to engage in far-reaching speculations. Some perceive the workings of the hand of God; others speak of a sufficient number of universes so that the law of large numbers would permit one among them (namely ours) to manifest the observed configuration of physical constants.

But the most widely discussed attempt at explanation is the 'anthropic cosmological principle.' It has been formulated in a number of ways.[15] The Weak Anthropic Principle (WAP) asserts that the observed values of all physical and cosmological constants are restricted by the requirement that there should be sites where carbon-based life can evolve within the known time frames. The Strong Anthropic Principle (SAP) goes on to claim that the universe *must* possess the properties that permit intelligent life to develop at some stage in its evolution. The Participatory Anthropic Principle (PAP) suggests in turn that intelligent life and observation are essential to the existence of the universe, since observers are required to bring the universe into being (a tenet derived from the phenomenalist interpretation of quantum theory); while the Final Anthropic Principle (FAP) postulates that intelligent information processing must come into existence in the universe, and when it does, it never dies out.

The strengths and weaknesses of these formulations have been widely discussed in recent years, and we need not add to this literature. Suffice it to note, following Errol Harris, that both the WAP and the SAP are self-evident truisms, because the fact that they are postulated in the first place is clear evidence that intelligent life exists.[16] From this it follows that the physical conditions for its existence, whatever they may be, must be given in the universe. Alternative universes in which conditions for the evolution of intelligent life are not given could not be observed, hence for the anthropic principle they are not merely speculative—they are inadmissible. The PAP, in turn, makes the physical conditions required for the evolution of intelligent life dependent on the activity of observation to which they gave rise, whereas in any realistic view the act (and fact) of observation must be held dependent on prior physical conditions. And the FAP is highly speculative: it predicts the development of information processing technologies to an extent that is as yet far from realized.

The various formulations of the Anthropic Principle, the same as theological and multi-universe concepts, owe their existence to the fact that the standard account of the emergence of order and complexity in the universe is woefully inadequate. It is beset by the problem of chance—a problem that surfaces in cosmology in the form of a haphazard increase of global entropy.

The standard argument takes off from the consideration that in an evolving cosmos, irreversible processes lead inevitably to the increase of global entropy. While local reversals of the sign of entropy change are possible, they must always be compensated by global entropy change with the opposite sign. Local entropy decrease is likely, because in an inhomogeneous expanding universe matter cannot assume an equilibrium distribution all at once. Dispersed matter searches haphazardly for equilibrium, with particles colliding and forming more complex particles and clumps of particles. The resulting configurations collide further and produce friction, further slowing the global entropy-increase. Complex configurations of particles, such as those required for life, are a random by-product of the overall entropy-oriented process. .(Mind and consciousness would be no more, of course, than accidental epi-phenomena.)

The above explanation overcomes the problem of incompatibility between the second law of thermodynamics and the local build-up of complexity by invoking the element of pure chance. Yet in regard to the orders that arise in nature, this is not sufficient: the iterations of a random-event mediated process would take far too long to generate the observed orders. Moreover even if the iterations were governed by nonrandom algorithms, the resulting process, though it could generate significant orders, could not set its own parameters.

In regard to the phenomena of precisely tuned universal constants, we are left with three alternatives. Either the constants were set by a mysterious agency at the beginning of the process; or they were set by pure chance; or the process itself has set them. Not being in favor either of mysterious forces or of extreme improbabilities, we choose the third alternative.

Non-Big-Bang cosmologies

The values of the universal constants must have been set already as the Big Bang occurred; thus it is not easy to see how the standard 'BB' scenario could account for their fine-tuning. There are, however, viable alternatives to this scenario.

The BB scenario, though widely applauded, is in trouble. The range of problems encountered by it includes a failure to explain galaxy formation (and even to find the 'fingerprints' of it in the cosmic background radiation); to identify the 'missing mass' in the universe; to solve the age problem of galaxies and stars and of the universe as a whole; and to explain how the inflationary process could have been switched *on* as well as *off*. The standard scenario also fails to clarify the origins of the universe prior to the Big Bang, and its destiny after the disappearance of matter in expanding or collapsing galaxies — in addition to explaining the observed ratio of baryons to photons and the mass of the top quark as well as of the Higgs boson.

Adherents of the BB scenario point out that many of these questions are not detrimental to their model—for example, whatever triggered the Big Bang, an inflationary epoch or whatever else, is not intrinsic to it. Also the question as to where the universe is heading (recollapse, infinite expansion or something still different) is part of the uncertain parameters with which cosmologists must learn to live. The standard scenario, they point out, yields nevertheless a set of meaningful interpretations and successful predictions, and these substantially outnumber its failures. Some cosmologists go as far as to assert that there is no known alternative cosmology that would take account of the full range of observational and experimental evidence.[17] Yet a handful of alternative cosmologies have now been elaborated.

The classical alternative to the BB scenario was the steady-state cosmology. This was a dominant view until about 1965, from which time the Big-Bang scenario began to eclipse it. In its original formulation the concept assumed a homogenous expanding universe that is not significantly evolving: matter is continually created so as to preserve a steady mean density.

The steady state concept goes back to an idea of Jeans'. In 1929 Jeans wrote, "The type of conjecture which presents itself somewhat insistently is that the center of the nebulae are of the nature of 'singular points' at which matter is pouring into our universe from some other, and entirely extraneous dimension, so that, to a denizen of our universe, they appear as points at which matter is being continuously created."[18] In the

1960s H.C. Arp and Fred Hoyle developed this notion into the modern form of the steady-state cosmology, replacing the idea of 'matter continuously pouring into the universe from an entirely extraneous dimension' with that of matter-creation *within* it. But the creation of matter in the universe could not occur at a constant rate; this would conflict with the presence of quasars and other active galactic nuclei, the discovery of the microwave background radiation, and the recognition that the basic constituents of matter are quarks and leptons.

In current (1993 and later) versions of the Quasi-Steady State Cosmology, Hoyle, Burbidge and Narlikar show that matter-creation occurs in bursts in the strong gravitational fields associated with dense aggregates of preexisting matter, in the nuclei of galaxies, for example.[19] Matter is created in 'little big bangs' of the order of about 10^{16} solar mass; thus there is no need to explain how the Big-Bang-type inflationary process could have been switched on, and again off. Matter-creation occurs through a scalar 'C' (creation) field of negative energy, of which the value is a function of space-time. The overall rate of matter-creation is determined by the square of the time-derivative of C, averaged over the universe.

As these dense gravitational events drive the universe's expansion, the rate of expansion is not constant but varies secularly with changes in the number and mass of the creation centers. The universe is generally expanding, with a superposed oscillation period of 40 billion years. It is at these intervals that matter-creation is concentrated, in a cycle that stretches back infinitely in time, to an epoch when the scale of the universe required an oscillatory minimum. The most recent burst of major matter-creation turns out to have occurred about 14 billion years ago, in good agreement with conservative Big Bang model estimates.

The QSSC can account for the general outpouring of mass and energy from a wide range of extragalactic objects, ranging from protogalaxies through high-energy events such as radio galaxies and quasi-stellar objects, down to small scale events in the nucleus of our own galaxy. It can also account for the observed age and distribution of galaxies and the synthesis of a variety of observed matter-particles. The random thermal energy associated with matter radiating from the creation centers is in agreement with the requirements for explaining the genesis of the cosmic microwave background.

As some other recent models, the QSSC constitutes a multicyclic cosmology. The standard generation length ($\sim H_0^{-1}$, according to Hoyle *et al.*), which makes for a relatively steady envelope, [...] admits 40-billion year exceptions. [...] roximates the Big Bang, producing [...] of the background radiation and the [...] viewed as including the Big Bang

cycles, matter from the current cycle [...] galaxies created in the current 14- [...] that were created in earlier cycles. [...] from the immediately foregoing cycle [...] 5.166, compared with the galaxies of [...] 4.86. The major part of the cosmic [...] of radiation from the previous cycles: [...] observed properties of the microwave tons in the universe were created about

[...] of Prigogine, Geheniau, Gunzig, and [...] the large-scale geometry of spacetime creates a reservoir of negative energy [...] which gravitating matter extracts positive energy; in the unstable vacuum the extracted energy is then transferred to virtual particles. Thereby energetic conditions are created for the synthesis of matter particles. There is a constant and balanced interaction between matter in the large-scale structures of the universe and the quantum vacuum. In each cycle of the universe particles of matter are created in the vacuum through the energy generated by the particles that were synthesized in the previous cycle. The positive energy going into the synthesis of matter constantly and precisely compensates the negative energy generated by the curvature of spacetime owing to the gravitational attraction of preexisting matter. Thus the theory describes a perpetual mill for the creation of matter. The more matter-particles have been generated, the more negative energy is produced, transferred as positive energy to the synthesis of still more particles.

New universal cycles arise because expanding galaxies 'dilute' the vacuum, and as dilution reaches a critical value it creates an instability. At that point the vacuum transits into the inflationary condition, and that phase in turn transits to the more sedate expansionary mode of the currently observed universe. The phase-transitions are governed by the creation and evaporation of minute black holes of the order of 50 times Planck mass m (where m equals $2.17671 \cdot 10^{-5}$ grams), according to the process described by Hawking. The evaporation time of these mini-black holes turns out to be 10^{-37} second—exactly the time required for the inflationary-phase by independently formulated theories.

Here the genesis of each cycle involves three stages separated by two phase-transitions. The first stage is the creation of an unstable Minkowski vacuum by the negative-energy feedback from the curvature of spacetime due to the universe's large-scale structures. The instability creates the phase transition to the inflation typical of the De Sitter universe: this is the second stage. During the inflationary stage the minute black holes created in the first phase-transition evaporate, producing a supercooling effect which, as the second phase-transition, drives the De Sitter universe into the third stage: the spatially homogeneous, isotropic and geometrically expanding Robertson-Walker universe.

The Prigogine, Geheniau *et al.* multicyclic cosmology is self-consistent: its phase transitions depend directly on the values of three constants: the speed of light c, Planck's quantum constant h, and the gravitational constant G. Because the vacuum is unstable in the presence of gravitational interaction, matter and vacuum form a self-generating feedback loop. The critical matter-triggered instability causes the vacuum to transit to the inflationary mode, and that mode marks the beginning of another era of matter synthesis. As a result the universe we observe was not created out of an unexamined pre-existing cosmic background, but arose as a new cycle within an already existing quantum vacuum.

The multicyclic QVI scenario

The cosmological alternative required for a meaningful explanation of the observed correlation of the universal constants is obviously of the multicyclic variety. Within multiple cycles of cosmic evolution, the parameters of the process—the constants themselves—can evolve. This, tenet, as we shall see, is consistent with the concept of quantum/ vacuum interaction.

The QVI scheme calls for the interference patterns of scalar wavefronts to superpose in the vacuum and form Schrödinger-type holograms. The state-spaces of quanta and the $3n$-dimensional configuration-spaces of multiquantal macroscale systems are encoded in this hologram, and matching quanta and multiquantal systems decode this information in the form of inverse Fourier transforms. In the resulting cosmological cycle information is created in the spacetime domain of quanta and is conserved in the spectral domain of the vacuum. Information-creation and -conservation exhibit a temporal asymmetry: the information-creating domain of quanta is limited in time (as quanta, and matter-energy systems built of quanta, evolve and devolve in space and time), whereas the information-conserving domain is temporally unlimited. (The time-path of particles is finite, whereas that of the vacuum's Schrödinger holograms is—as far as one can tell—infinite).

The above temporal asymmetry is significant if in the cosmos matter is periodically created. When matter is created, the particles that are synthesized emerge from, and interact with, the quantum vacuum. Assuming that the topology of the vacuum results from the prior evolution of matter in space and time, the particles created in a cosmic region are 'informed' by the Fourier transforms of the state-spaces assumed by particles in the prior universal cycles.

In a distributed and multidimensional form, the Fourier transform of the state- and configuration-spaces of all events in space and time is encoded in the vacuum. But this 'wave-function of the universe' cannot be available simultaneously in all spacetime regions: scalar waveform availability must be limited by the Weyl spacetime cone of causal propagations, as specified by the variable velocity of scalar wave-propagation in the ZPF. This means that in any given cosmic region some, but not all, elements of the wave-function of the universe fall within the spacetime cone. As some elements of the waveform traces of matter in prior cycles will be present in whatever region a burst of matter-creation may occur, the vacuum can be said to transfer 'information' (in the active sense of 'in-formation') from one cycle to the next.

It may be helpful in this context to return to the vessel/sea metaphor suggested in Chapter 2. Here we envisage a finite but enduring sea in which a perturbation—a bubble resulting from some pressure—creates concentric waves that propagate from the center toward the periphery. The receding wavefronts exercise a reverse pressure that propagates from the periphery to the center, so that eventually there is another bubble creating another set of spreading wavefronts. As these, too, move outward they feed back their own reverse pressure which then catalyzes a third bubble—and so on.

We allow that the sea can conserve the traces of the wavefronts that were generated in it. As each wavefront created by a bubble passes outward, the surface does not die back to smoothness but remains subtly modulated. As a result the wavefronts created by the next bubble become modulated by the enduring topography of the surface. When the successive wavefronts interact with the traces of the prior wavefronts, they add further wave-patterns to that surface. The more wavefronts pass over the surface, the more it becomes modulated, and the more it modulates the successive wavefronts. As the process repeats, the existing and the preexisting patterns become increasing coherent and consistent. Vessels that come and go over the surface are 'in-formed' not only with their own quasi-instantaneously spreading wakes, but with the progressively more numerous and coherent wavepatterns resulting from the interfering wakes of the previously passing vessels.

The tenet of multicyclic information transfer exhibits a good fit with current ideas regarding the mechanisms of matter- (or 'universe'-) creation. Alan Guth and other Big Bang cosmologists have shown that the inflation through which the universe we now observe came into being was highly, but not completely, uniform. To account for the gravitational clumping of matter into galactic clusters—and then into galaxies and stellar systems—we must assume that the initial radiation field contained minute variations that were enlarged during the inflationary period. Theory holds that these variations gave rise to scale-invariant fluctuations (that is, fluctuations with an amplitude independent of physical size) in the gravitational field; and the resulting fluctuations, through the Sachs-Wolfe effect, produced measurable temperature differences in the cosmic background radiation. This is substantiated by empirical evidence: COBE's differential microwave radiometer (DMR) found scale-invariant fluctuations in the temperature of the cosmic background. DMR's instrumentation was able to detect background radiation on angular scales greater than seven degrees; and those it actually detected were of the order of ten degrees or more.

But from where did these scale-invariant fluctuations come? Although there is no universally accepted answer, the most widely held view, outlined among others by Guth, Hawking, and Alexei Starobinsky, is that they were produced by primordial fluctuations of the quantum vacuum. (The alternative concept traces them to the phase-transitions that followed the Big Bang, as the universe passed from higher to lower energy states.) In the prevalent view the initial quantum fluctuations expanded by a factor of at least 60 during inflation; they then introduced the inhomogeneities in the radiation field that gave rise to the gravitational clumping of matter in spacetime. BB-theorists are obliged to view the fluctuations as independent variables: on the Big Bang scenario the universe has no history prior to inflation. However, on a multicyclic scenario the universe does have a prehistory.

The QVI scenario links the primordial quantum fluctuations informing inflation to that prehistory: the fluctuations are dependent variables within a self-consistent iterating process.

In the view represented here the primordial vacuum fluctuations contain the wave-transform of the particles that fall within the spacetime cone of scalar propagation in the matter-creation region of the vacuum. This makes the successive cycles of the universe elements in a continuously interlinked non-Markovian chain. The evolution of micro- and macroscale systems in each cycle creates the initial fluctuations that determine in turn the 'graininess' of the inflation that produces the next cycle's large-scale structures.

The trans-cyclic transfer of information does not cease with the co-determination of the dimensions of the large-scale structures of the universe: as these structures evolve they create physico-chemical templates suitable for the evolution of complex supramolecular systems. The systems that come about in one cycle continually interact with the holographic information created in the vacuum by the systems that evolved in the foregoing cycles. Thus we get a significant probability that in the successive cycles large-scale structures should appear in which some stars have planets capable of supporting life, and that on some of the planets complex multimolecular systems typical of life should evolve.

For the multicyclic QVI scenario the universe is spatially finite but temporally infinite. It evolves without end over a trans-cyclic topological manifold that progressively constrains random processes into forms of order. The manifold tunes the universal constants to the micro- and macroscale systems that evolve in space and time.

Because at present the universal constants are finely tuned, we must assume that the current cycle has not been the first. For the multicyclic QVI scenario the evolution that the standard scenario views as the complete lifetime of the universe is but one of an infinite sequence of evolutionary cycles. The universe is a temporally infinite, self-renewing, and strongly interacting system, progressively adapting itself to the evolution of complexity—and hence to the evolution of life and, perhaps, also of mind.

References

1. John Archibald Wheeler, 'Bits, quanta, meaning,' in *Problems of Theoretical Physics*, A. Giovannini, F. Mancini, and M. Marinaro (eds.). University of Salerno Press, Salerno 1984.

2. Jean Staune, 'La révolution quantique et ses conséquences sur notre vision du monde'. *3e Millénaire*, **15** (1989).

3. Eugene Wigner, *The Scientist Speculates*, I.J.Good (ed.), Heinemann, London 1961.

4. R.G. Jahn and B.J. Dunne, 'On the quantum mechanics of consciousness, with application to anomalous phenomena.' *Foundations of Physics*, **16,** 8 (1986).

5. H. Everett, *Rev. Mod. Physics*, 29. 1957.

6. Paul Dirac, in *Proc. Einstein Centennial Symposium*, Jerusalem 1979.

7. David A. Chalmers, personal communication 22 September 1994.

8. A. Aspect, P. Grangier, and G. Roger, in *Phys. Rev. Lett.* **49**.9 (1982).

9. O. Costa de Beauregard, *Le Temps Déployé*, Editions du Rocher, Monte Carlo 1988.

10. Huw Price, 'A neglected route to realism about quantum mechanics', *Mind*, **103** (July 1994).

11. Wheeler, 'Bits, quanta, meaning,' *op. cit.*

12. David Kleppner, Michael Lettmann and Myron Zimermann, 'Highly excited atoms,' *Scientific American* (May 1981).

13. E. Del Giudice, S. Doglia, M. Milani, C.W. Smith and G. Vitiello, 'Magnetic flux quantization and Josephson behaviour in living systems.' *Physica Scripta* **40** (1989), 786-791.

14. Wheeler, Quantum Cosmology, *World Science*, L.Z. Fang and R. Ruffini (eds.), Singapore 1987.

15. John D. Barrow and Frank J. Tipler, *The Anthropic Cosmological Principle*, Oxford University Press, London and New York 1986.

16. Errol E. Harris, 'The universe in the light of contemporary scientific developments,' in M. Kafatos (ed.), *Bell's Theorem, Quantum Theory and Conceptions of the Universe*. Kluwer Academic Publishers, New York 1989;
 —, *Cosmos and Anthropos,* Humanities Press, New York 1991.

17. P.J.E. Peebles, D.N. Schramm, E.L. Turner and P.G. Kron, 'The case for the relativistic hot Big Bang cosmology,' *Natur*, **352** (29 August 1991).

18. J.H. Jeans, *Astronomy and Cosmogony*, Cambridge University Press, Cambridge, 1929.

19. F. Hoyle, G. Burbidge and J.V. Narlikar, 'A Quasi-steady state cosmology model with creation of matter,' *The Astrophysical Journal*, **410** (20 June 1993).

20. E.Gunzig, J.Geheniau and, I. Prigogine, 'Entropy and Cosmology,' *Nature* **330**, 6149 (December 1987);
I. Prigogine, J. Geheniau, E. Gunzig, and P. Nardone, 'Thermodynamics of Cosmological Matter Creation,' *Proceedings of the National Academy of Sciences, USA*, **85** (1988).

Chapter Four

Q V I IN BIOLOGY

The facts we confront in the biological world are evident and prima facie uncontroversial. We are faced with, and indeed immersed in, a vast nested hierarchy of systems within systems. Planetary ecologies, the largest of the terrestrial systems, consist of continental and subcontinental ecologies which in turn consist of populations of organisms within particular physical, chemical and biological environments. The individual organisms that inhabit these ecologies are made up of cells, cells are made up of proteins, proteins consist of a combination of molecular groups, molecules consist of atoms, and atoms are configurations of nucleonic and electronic quanta.

Darwinian and neo-Darwinian theories insist that this nested hierarchy has evolved by the natural selection of random mutations. The mainstream theories uphold the assumption of randomness of the germline's variations and maintain Weismann's doctrine of the separation between soma and germline. The latter is usually interpreted to mean that physiological interactions with the environment during the lifetime of the organism do not have heritable effects—they do not produce changes in germline DNA. In turn, the randomness of mutations is to ensure that genetic changes in the germline are influenced neither by the state of the phenotype nor by conditions in the environment. Adaptive evolution, classical Darwinists tell us, proceeds by an *a posteriori* selection of randomly produced genetic variants that happen to 'fit' particular environments. However, as Ho, among others points out, these assumptions—the randomness of mutations and the insulation of genes from environmental influence—have both been falsified by empirical findings in molecular genetics.[1] Why this is so can be briefly stated.

Problems of the synthetic theory

To begin with, sequences of DNA in the chromosomes of cells cannot by themselves account for the differentiation and organization of a great variety of cells under a great variety of conditions. François Jacob noted that what distinguishes a butterfly from a lion, a hen from a fly, or a worm from a whale is much less a difference in chemical constituents than in the organization and distribution of these constituents.[2] Among vertebrates, for example, the chemistry is the same; differences between different species cannot be traced to structure: they are a matter of regulation. Minor changes in the regulatory circuits during the development of the embryo can radically affect the final result: a substantially different animal can be produced just by changing the growth rate of different tissues, or the time of synthesis of various proteins. DNA, however, is the same in each cell of the developing organism. Even if it interacts in a differentiated way with the milieu of the cell, it is difficult to see how genetic information alone could be responsible for the delicate and highly specific tuning of organic regulation. Even to assemble a single cell, the information contained in DNA is not sufficient: localization in the cell's assembly depends on processes other than those involving DNA transcription—investigators have found that an existing set of cellular structures is needed to direct the assembly and organization of the components of the new set of structures. This discovery has been said to call for an 'epigenetic revolution' just as exciting as the previous molecular genetic revolution.[3]

The classical doctrine fails in regard to the insulation of the genome. Alterations occurring in DNA both during development and in evolution prove to be sufficiently large to have prompted some molecular geneticists to coin the phrase 'the fluid genome'. DNA appears to be structurally and functionally as flexible as the rest of the organism. Recent findings suggest that, even if some nucleotide changes in the DNA are fortuitous, the resulting variations in the organism are not random but occur within the context of a highly structured epigenetic system. The dynamical structure of this system shapes the variations in the germline of species in such a way that variations result not only from the random variability of the genome, but also from environmental factors that act on the genome's variability.

Many germline changes are found to have been produced as the result of specific environmental perturbations; the resulting changes are then passed on to subsequent generations. Such basically Lamarckian evolution is exhibited by directed genomic changes occurring in flax and other plants following treatment with fertilizers, and by various insect species that, exposed to insecticides, produce heritable amplifications of specific genes that detoxify the chemicals and create genetic resistance to the toxins. The ability of bacteria to mutate under particularly stressful conditions is perhaps the most striking evidence for environmentally influenced genetic variation. 'Starving' bacteria can selectively mutate the exact strand of information in the precise genes in just such a way as to metabolize the food it is fed.[4] But how bacteria could detect which information in which gene is incorrect and what is the correction necessary to rectify the condition, is beyond the classical theory.

The survival of species in a changing environment creates difficulties of a different kind for the Darwinists' 'synthetic theory'. From time to time, in order to survive, species have to shift from one niche to another. On Darwinian premises it is not clear how they could do so. A random sequence of mutations in the genetic code could not have produced the leaps that hallmark the emergence of new species—the mutations involved in speciation are 'systemic': highly coordinated and massively adapted to the current or emerging conditions of survival and reproduction. If evolution relied entirely on random mutations, the mutants that happened to be adapted to a new niche would not survive long enough for the species to reach that niche. Given that they are bound to be less adapted to the current niche than the non-mutant lineages, they would soon be eliminated by natural selection.

In the newer versions of the synthetic theory chance-driven processes of continuous small-scale adaptation—termed 'phylogenetic incrementalism'—are questioned. About a hundred years after the publication of Darwin's *Origin of Species* (where Darwin declared: `Natural selection...can produce no great or sudden modifications; it can act only by short and slow steps'), Jay Gould and Niles Eldredge formulated the theory of evolutionary leaps, known as the theory of 'punctuated equilibria'.[5]

Gould and Eldredge sum up the contrast between the classical evolutionary theory and the theory of punctuated equilibria in reference to the transformation that occurs as new species emerge. For Darwinian theory such transformation:

— concerns an ancestral population that evolves into its descendants by means of gradual modifications;
— is even and slow;
— involves large numbers, usually the entire ancestral population;
— and occurs over all or a large part of the geographic range of the ancestral species.

If so, the fossil record should consist of a long sequence of continuous, insensibly graded intermediate forms linking ancestor and descendant, with morphological breaks in this sequence due only to imperfections in the geological record. This, however, does not seem to be the case. The record, as Gould and Eldredge observe, shows that 'speciation' is a rapid process: new species burst on the scene within time periods of the order of 5,000 to 50,000 years. Not only individual species but entire genera make their appearance in sudden epochs of creativity. For example, the Cambrian explosion brought forth in the span of a few million years most of the invertebrate species that now populate the Earth.

The fossil record proves to be discontinuous, and the 'missing links' are more likely to be due to nature than to imperfect knowledge. The theory of punctuated equilibria asserts that new species:

— arise following the splitting of lineages;
— develop rapidly;
— come about in a small sub-population of the ancestral form;
— and originate in a small part of the ancestral species' geographic extent, in an isolated area at the periphery of the range.[6]

Evolution, punctuated equilibria theory claims, acts on species and populations and not only—or even mainly—on individual reproducers. Individual variations do not contribute significantly to the emergence of new species; the classical Darwinian mechanism works mostly to adapt individuals to their existing niches. When the milieu changes and the existing niches disappear, species tend to die out—the 'peripheral isolates' invade the centers of dominance and take over as the new dominant species.

But the rejection of gradualism in favor of evolutionary bursts does not remove the factor of randomness in evolution; the Gould-Eldredge theory merely shifts the element of chance from individual survivors and reproducers to an entire hierarchy that includes species and populations. As Gould points out, the new theory advances above all two sorts of proposal: a widened role for non-adaptation and for chance as a source of evolutionary change; and the construction of a hierarchical concept based on the interaction of selective (and other) forces at numerous levels from genes to entire population groups, rather than almost exclusively upon selection among organisms.[7]

Chance, however, even in the currently conceptualized form of 'non-adaptive and non-selective drift,' remains a questionable factor in the processes of evolution. How could evolution by chance produce the kind of changes in the genome that could assure the viability of a new species? It is not enough for mutations to produce one or a few positive changes in the organism; they must produce a full set. The evolution of feathers, for example, does not produce a reptile that can fly: radical changes in bone structure and musculature are also required, along with a faster metabolism to power sustained flight. Each innovation by itself is not likely to offer evolutionary advantage; on the contrary, it is likely to be unfit and hence eliminated.

The calculation of statistical probabilities argues against the assumption that evolution could have proceeded by a random stepwise elaboration of the genetic code of the surviving species. If a significant evolutionary step, such as the development of an eye or another organ, is to have occurred in living nature, strictly sequenced mutational steps must have been produced. However, R.J. Gilson points out, the probability that just 12 such steps should appear in strict numerical sequence in 12 consecutive trials (for example, in 12 consecutive throws of the die) is 1 in 12^{12}. This makes the probability that any mutation should spontaneously produce a viable new species astronomically small.[8]

We must consider, of course, that evolution is not a single-throw affair; it proceeds by repeated trials and errors, and repetition increases the probability that a given sequence of steps should come about. For example, if the steps required for producing a viable new species numbered merely 12, and if they were repeated one million times, the probability of their occurrence would be 'merely' 10^{-7}. But this underestimates the complexity of the required mutations. Significant evolutionary developments, such as the creation of an eye, or of the capacity to fly, number far more than 12 steps. If we take the number of required steps to be but one hundred, the probability of their random occurrence in the required sequence rises to 100^{-100} (that is, 10^{-200}).

Unlike in the laboratory, where powerful computers modelling interactive processes can carry out billions of operations per second, the number of iterations of the evolutionary process is subject to strict limitations in nature. There cannot be an infinite number of trials in a finite time span: the maximum number of trials we can reasonably assume evolution to have produced is 10^{11}—this assumes that in the 600 million years that separates us from the Cambrian, a mutation occurred every few hours. The combination of the probability of the one-time occurrence of one hundred steps in rigorous sequence (10^{-200}) with the 10^{-11} assumed number of trials gives us a probability of ($10^{-200}-10^{-11}$) = 10^{-189}. This still astronomically small probability is made even smaller by the fact that:

(1) mutational trials are not likely to have been produced incessantly one every few hours for the past six hundred million years, and

(2), in nature not one hundred, but many thousands of highly coordinated steps are required to create a viable mutant.

It is with good reason that M. Schutzenberger noted that one would need an almost blind faith in Darwinian theory to believe that chance alone could have produced in the line of birds all the modifications needed to make them high-performing flying machines, or that random mutations would have led to the line of mammals after the extinction of the dinosaurs—given that mammals are a long way from dinosaurs along the axis that conduces from fish to reptiles. Evolution, Schutzenberger concluded, contradicts categorically Gould's thesis on chance.[9] Giuseppe Sermonti concurred: it is hardly credible, he said, that small random mutations and natural selection could have produced a dinosaur from an amoeba.[10]

Major evolutionary novelties are not likely to have resulted from the gradual accumulation of minor changes: the variational possibilities are too vast and the observed leaps between species too great for random variation in species genotypes to explain the observed course of evolution. We must assume that continued fitness is not the result of random mutations exposed to the test of natural selection and macroevolution is more than the simple sum of the set of randomly produced and naturally selected micro-evolutionary modifications. While natural selection plays a role in evolution (variations that are distinctly disadvantageous do not persist, and this contributes to the observed fit between organism and environment), natural selection could not even have led to the bisexual mode of reproduction. Such a mechanism, Saunders argued, though offering an obvious long-term advantage (the more rapid spread of advantageous mutations) involves an equally obvious short-term disadvantage: the reduced average number of descendants due to males failing to produce offspring.[11] Hence selection in nature is a negative rather than a creative factor: it weeds out the unfit mutants, but it does not ensure that there should also be mutants that are fit. If so, genetic variation in a population is not a matter of selection from among fully random mutations. Instead, as Ho said, "the dynamic structure of the epigenetic system...organically 'selects' the 'response' or action that is appropriate."[12] Organism and environment, she asserted, are closely interconnected, from the sociocultural level right down to genomic DNA.

Though neo-Darwinists now contemplate and occasionally affirm the effect of environmental systems on the evolution of species, they cannot give a detailed account of it.[13] How systems at a high level of one hierarchy, such as populations, communities and regional biotal systems, could interact with systems at a low level of another hierarchy, such as genes, organisms, and the genetic information encoded in the genome, is not explained on Darwinian premises alone.

Interaction effects on the organism

We can now outline the thesis that mitigates the difficulties of the synthetic theory by taking into account the emerging evidence for the closer than suspected connection between the genotype and the phenotype, as well as the between the whole organism and its environment.

The in-formation of the organism

It now appears that the organism is linked to its own species-specific morphology as well as to the dynamic systems and configurations that embed it in its environment. Even though in classical Darwinism mutations were said to be internal to the organism and selection pressures external, in post-Darwinian versions of the synthetic theory the distinction between 'inside' and 'outside' became fuzzy. Richard Lewontin asserted that the evolutionary process links organism and environment both intrinsically and inseparably, and Susan Oyama identified the unit of evolution as the integral 'system of development.' It includes the nucleus of the cell together with the cellular structures and the innumerable extracellular influences; it also includes the full organism together with all aspects of its environment that influence its development, from the chemical substances the embryo draws from the mother's body to the social relations to which the organism is exposed through the various stages of its growth and maturation.

In the emerging post-Darwinian view the organism is an integrated whole that, in Ho's words, is 'transparent' to its environment.[18] Such transparency cannot be accounted for in physical and biochemical terms: it involves a remarkably precise and complete set of signals linking the organism with its adaptive landscape. This points toward a multidimensional signal-transmitting field that would in-form the organism.

As already remarked, field concepts have been explored in biology since Alexander Gurwitsch postulated his concept of the morphogenetic field in the 1920s. Gurwitsch was led to the field concept by noting that in embryogenesis the role of individual cells is determined neither by their own properties, nor by their relations to neighboring cells, but by a factor that involves the entire self-organizing system. He postulated a system-wide 'force field' generated by the force-fields of individual cells. Although at first Gurwitch claimed that these fields are non-material, he later allowed that the concept could be translated into the language of physics.

The early biofield concept was elaborated by several biologists, including N.K. Kolchov in Russia, Ervin Bauer in Hungary and Paul Weiss in Vienna. Conrad Waddington and René Thom connected generative processes with geometrical forms by conceiving of the field as consisting of zones of structural stability. Brian Goodwin has given numerous demonstration that biofields are associated with growth processes in plants and animals. In his view the field is the basic unit of organic form; molecules and cells are merely units of composition.[19]

Fields can account for a variety of generative and regenerative processes in biology; it is likely to become a central notion in post-Darwinian theory. Current evidence points to a radiation field acting on matter in general and on living organisms in particular. Del Giudice and Preparata have shown that above a critical threshold of density the ambient radiation field influences processes in substances such as water, as well as in the tissues of living organisms (in terms of total weight the latter are about 80 percent water). Ordinarily molecules and other microscopic components interact through short-range static forces (H-bonds, London forces, van der Waals forces) while the contribution of the electromagnetic field to their interaction is small and thus generally neglected. However, when the number of components is large with respect to a quantity dependent on the wavelength λ of the interaction (i.e., when the average distance δ among the microscopic components becomes negligible with regard to λ [$\delta \ll 1$]), then the fundamental configuration of matter and EM field becomes unstable. The system settles into a new configuration, called the 'super-radiant vacuum', where most of the components are kept in phase by a time-oscillating electromagnetic field that is constant over a coherence domain extending over $1/2\,\lambda$.[20]

If the oscillations of the EM field responsible for the coherence domain are mediated by scalars generated by the motion of quanta, living tissue (as well as water and other substances) is 'in-formed' with the interference patterns that code their state- and configuration-space. The initial effect, which may be extremely subtle, is registered through the input-sensitivity of the organism and is amplified by its chaos dynamics. As a result the organism is in-formed with the wave-function of its own species-specific morphology, as well as with the complex wave-function of the populations and ecosystems in which it finds itself. We can thus envisage, with Del Giudice, the possibility that in the macroscopic domain of a few hundred microns coherent interaction between the radiation field and sensitive components in living tissue generates ordered structures.[21] These structures may have a fundamental role in the organization of living matter.

The presumed effect of fields will be illustrated here in regard to two major instances of organic generation and regeneration: embryogenesis, and neurulation (the genesis of the nervous system).

Embryogenesis constitutes a complex and in part chaotic system. Enormously intricate processes are involved in cell-division and differentiation, calling for detailed and accurate regulation. Such regulation cannot be mapped by stable attractors, and it is not likely to be governed by genetic information alone. Regulation is more likely to result from an interaction between DNA-coded cells and an 'epigenetic' landscape, with numerous chaotic—but not random—features.

In the context of embryogenesis, the concept of epigenetic landscape was originally intended by Waddington to refer to the complex milieu formed by genes, environment and the organization of the morphogenetic field. This milieu defines the attractors available to the embryo, enabling it to choose the chreods (dynamic pathways) required for its development. In the present interpretation, it is the holofield of the quantum vacuum that acts as the morphogenetic field. Given that the cellular growth-system is significantly chaotic—hence ultrasensitive—the information encoded at the vacuum level can be called upon to account for the precise regulation of the growth and differentiation of the embryo. In this view embryogenesis is the outcome of an interaction between DNA-coded cells, the biochemistry of the womb, and the holographically pattern-conserving and transmitting scalar-mediated electromagnetic spectrum of the quantum vacuum.

In reviewing this tenet we should keep in mind that the cells that make up the developing embryo constitute a dynamic configuration of atoms and molecules with pronounced chaos dynamics. As all chaotic systems, this dynamically indeterminate ensemble is highly sensitive to minute variations in its internal and external parameters. Unmeasurably fine fluctuations reaching the system may produce measurable—in fact, decisive—effects on its evolution. The QVI scheme suggests that this ultrasensitive growth-system is in continuous interaction with the multidimensional waveforms translated into the vacuum holofield by generations of organisms of a particular species. The embryo's chaos dynamics registers this minute input and translates it into a precisely biased selection of evolutionary trajectories. Thus feedback from field can be seen as orienting the interplay of the pathways of differentiation, governing the growth-rate of various tissues and the time of synthesis of different proteins. The embryo's development along pathways consistent with the morphology of its species surfaces as the result of this interactive field guidance—the consequence of the embryo's 'sacred dance' with the ambient biofield.

In the case of neurulation the requirement for field-guidance is just as pronounced. The human cortex contains about ten billion neurons forming one million billion connections. During its formation the neural tube produces precisely coordinated processes that yield this astoundingly complex cortical mantle. However, the way the six-layered mantle is generated does not follow any preestablished program. Neurons move within and among the layers with a high degree of freedom; many die before the structures they create would reach maturity. The dendrites they form at the points where they connect with other neurons assemble into complex arbors that show an overlap of up to 70 percent. It is impossible to determine from which neuron any given connection or synapse would have come. Moreover each time a neuron sends out axons to connect with other neurons, it branches in ways that are not predictable in reference to its prior state or connections.[22]

Neuronal connections differ in identical twins, and they differ from place to place in the same individual. Even in simple organisms, such as *Daphnia magna* (water fleas), genetically identical neuronal structures differ from individual to individual; while within a single organism (for example, a rabbit's cerebellum), there are no two structures that would repeat in exactly the same way. Because of these seemingly random variations, each brain or nervous system exhibits a neuronal pattern that is different in specific detail, even if the resulting functions are similar or nearly identical.

Edelman noted that the number of connections created in a neural system is far too great to be accounted for by any amount of genetic information. Also the way these connections are formed excludes reference to a morphological dynamics that would operate mechanically, as a computer or a Turing machine, on the basis of DNA-coded information; some other mode of organization is called for.[23] But that alternative mode cannot be, as Edelman maintains, a Darwinian selection of neural nets by trial and error. As shown both by Hoyle's example of the blind man working with a Rubik cube and by Edelman's own experiments with different versions of 'Darwin', a computerized neurulation-simulation program, the time required for a random process of this kind to produce the observed results exceeds the available timeframes. It is more reasonable to look for a field that would subtly but effectively guide the self-organizing dynamics of neurulation.

The observed flexible evolution of ten billion nerve cells into structures with one million billion connections requires a field that can orient large assemblies of neurons across random individual variations toward functionally equivalent solutions. If this process is to be achieved within admissible timeframes, the guiding field cannot be generated in the process itself (though the newly forming neuronal structures may constitute emerging local fields themselves): it must be there as a complete information base from initiation to completion. Constant vacuum-based interconnection satisfies this requirement. The vacuum's Schrödinger-hologram endures independently of the individual organism, is interactively form-orienting, and carries the full set of relevant information.

The above concept can shed light on the puzzles associated with morphological regeneration as well. The scalar-mediated vacuum holofield encodes the species-specific imprints of all organisms, and those that 'match' these imprints are constantly affected by them. The results, we have said, become manifest in chaotic states. Whenever the ultrasensitive dynamics of chaos come into play, the interaction-effect guides not only processes of generation, but also processes of regeneration. For example when, having been separated by a sieve, the cells of a marine sponge enter a chaotic state, the feedback of the species-specific $3n$-dimensional pattern of the sponge in its ensemble provides the subtle 'prompt' that the attractors of chaos inflate into an effective intercellular guide for the reassembly of the complete organism.

Basically the same dynamic is at work when the cells that make up the artificially removed lens in the eye of the newt reassemble, and when in certain species entire organs and limbs regenerate. In humans such quasi-miraculous forms of regeneration are not usual, but those that occur are remarkable enough. For example, when a finger is severed above the first joint, and if the wound is not surgically covered with skin tissue, the fingertip grows back fully, regenerating even the individual's unique fingerprint. Only an enduring field that codes the complete organism could orient cellular divisions to produce such a precisely tuned regenerative process.

In this study we identify the orienting biofield as the scalar-mediated ZPF. The Fourier transforms of the state-space of particles and the configuration-space of living systems are encoded and conveyed to systems in isomorphic spaces, in-forming them both with their own prior states, and with that of the larger systems in which they are integrated.

References

1. M.W. Ho, 'On not holding nature still: evolution by process, not by consequence,' in *Evolutionary Processes and Metaphors*, M.W.Ho and S.W. Fox (eds.), Wiley, London 1988, 117-144.

2. François Jacob, *The Logic of Life: A History of Heredity*, Pantheon, New York 1970. *Pattern Formation: The Problem of Assembly*, Karger, New York and Basel, 1991.

4. B.G. Hall, 'Evolution on a petri dish.' *Evolutionary Biology* **15** (1982).

5. Niles Eldredge and Stephen J. Gould, 'Punctuated equilibria: an alternative to phylogenetic gradualism,' *Models in Paleobiology*, edited by Schopf, Freeman, Cooper, San Francisco 1972;
Stephen J. Gould and Niles Eldredge, Punctuated equilibria: the tempo and mode of evolution reconsidered,' *Paleobiology* **3** (1977).

6. Niles Eldredge, *Time Frames: The Rethinking of Darwinian Evolution and the Theory of Punctuated Equilibria*, Simon & Schuster, New York 1985.

7. Stephen J. Gould, 'Irrelevance, submission and partnership: the changing role of paleontology in Darwin's three centennials, and a modest proposal for macroevolution,' D. Bendall, (ed.), *Evolution from Molecules to Men*, Cambridge University Press, Cambridge 1983.

8. R.J. Gilson, personal communication.

9. M. Schutzenberger, in *Figaro Magazine*, 26 October 1991.

10. Giuseppe Sermonti, in *Figaro Magazine, op.cit.*

11. Peter T. Saunders, 'Evolution without natural selection,' in *Journal of Theoretical Biology* (1993).

12. M.W. Ho, 'The role of action in evolution,' in *Cultural Dynamics*, **4** (1991), 336-54.

13. Niles Eldredge, *Unfinished Synthesis. Biological Hierarchies and Modern Evolutionary Thought*, Oxford University Press, Oxford 1985.

14. W. Grundler and F. Kaiser, 'Experimental evidence for coherent excitations correlated with cell growth.' *Nanobiology*, **1** (1992).

15. H. Fröhlich (ed.), *Biological Coherence and Response to External Stimuli*, Springer Verlag, Heidelberg 1988.

16. D.W. Duke and W.S. Pritchard (eds.), *Measuring Chaos in the Human Brain*, World Scientific, London and Singapore 1991;
B.H. Jansen and M.E. Brandt (eds.), *Nonlinear Dynamical Analysis of the EEG*, World Scientific, London and Singapore 1993.

17. R. Pool, 'Is it healthy to be chaotic?' *Scienc*, **243** (1989) 604-607.

18. M.W. Ho, 'The role of action in evolution,' *op. cit.*, 336-54.

19. Brian Goodwin, 'Development and evolution,' *Journal of Theoretical Biology*, **97** (1982);
—, 'Organisms and minds as organic forms,' *Leonardo*, **22**,1, (1989) pp.27-31.

20. E. Del Giudice, G. Preparata and G. Vitiello, 'Water as a free electric dipole laser,' *Physical Review Letters*, **61**,9 (1988).

21. *Ibid*, and E. Del Giudice, S. Doglia, M. Milani, and G. Vitiello, in F. Guttmann and H. Keyzer (eds.), *Modern Bioelectrochemistry*. Plenum, New York 1986;

22. Gerald M. Edelman, 'Morphology and Mind: is it possible to construct a perception machine?' *Frontier Perspectives*, **3**,2 (1993).

23. *Ibid.*

Chapter Five

Q V I IN THE COGNITIVE SCIENCES

Though the immediate objective of the scheme outlined in Chapter 2 is the linkage of physical and biological phenomena in a transdisciplinary unified theory, the empirical testing of its postulates can go beyond these disciplinary realms. After all, the sphere of mind and consciousness, though it requires specialized disciplines for detailed investigation, is an integral part of the living world, emerging in the context of the evolution of a highly complex and communicative biological species. It is logical to inquire whether this sphere would exhibit interconnections that can be consistently and parsimoniously ascribed to the same principles as those that apply to physical and biological phenomena.

Let us recall our principal findings. In the realm of physics, the scalar-mediated ZPF of the vacuum was found to interlink quanta, creating various phenomena of non-locality and coherence; in the cosmos, the interconnecting field was seen to synchronize the values of basic physical constants, producing the physicochemical basis for the evolution of complex molecular and supramolecular systems. In the biological realm, the vacuum-linkage of living systems was seen to interlink organisms with their species-specific morphology, skewing otherwise random bifurcations in evolutionary trajectories for consistency with the past as well as with the environment of the systems. We now inquire whether functionally analogous interconnections may apply to the minds and brains of human beings.*

In this Chapter we first examine evidence that suggests spontaneous interaction among human brain/minds, and then explore whether the QVI concept can offer a consistent and parsimonious explanation of such phenomena.

The interactive varieties of human experience

According to the Western scientific tradition, everything that we perceive comes to us through the senses. This is the basic tenet of classical empiricism, but it is not necessarily true. Even if the brain is the key organ in our traffic with the external world, that it would be limited to the flow of data transmitted by the five exteroceptive senses does not necessarily follow. While it is clear that the principal varieties of distance-perception originate either in the electromagnetic field (the visual data) or in the atmosphere (the acoustic data), this does not mean that the brain must be uniquely dependent on these and other sensory sources for information. Contents of consciousness that are not conveyed by eye, ear, and the other bodily senses are not necessarily inferences from sensory data, nor must they be products of the imagination.

ESP

Until recently, the investigation of extrasensory perception (ESP) was relegated to parapsychology, considered a nonscientific discipline. This attitude is no longer justified. Parapsychology has become an experimental science. While charlatans still persist at its fringes, quantitative methods now dominate the academic centers of teaching and research. Telepathy, for example, has been subjected to a wide array of experimental tests, and explanations in terms of hidden sensory cues, machine bias, cheating by subjects, and experimenter error or incompetence have all been considered. A number of statistically

---

* We shall not enter here into the perennial philosophical debate regarding the relationship between brain and mind. We recognize that these are qualitatively different phenomena but maintain that they are not ontologically distinct, but merely different aspects of the brain/mind system. Consequently whenever events occur in the mind, there are corresponding events in the brain—though, of course, not all events in the brain are reflected in the conscious mind. (For further details on this 'biperspectivism,' see Ervin Laszlo, *Introduction to Systems Philosophy*.[1])

significant results have survived nevertheless. We review here those strands of evidence that merit consideration as such bon fide items of observational and experimental data.

Evidence in regard to telepathy comes from anthropologists as well as parapsychologists. Anthropological evidence indicates that telepathy is common among so-called primitive people. In many tribal societies shamans seem able to communicate telepathically, using a variety of techniques to enter the altered states of consciousness that seem required for it, including solitude, concentration, fasting, as well as chanting, dancing, drumming, and the use of psychedelic herbs. Australian aborigines appear to be informed of the fate of family and friends, even when out of sensory communication range with them. A.P. Elkin noted that a man, far from his homeland, "will suddenly announce one day that his father is dead, that his wife has given birth to a child, or that there is some trouble in his country. He is so sure of his facts that he would return at once if he could."[2]

Anthropological data is largely anecdotal and unrepeatable. By contrast, evidence of a more reliable kind comes from laboratory research. Such research began systematically already in the 1930s, with J.B. Rhine's pioneering experiments at Duke University. In recent years the controls became increasingly rigorous, with physicists joining the design of the experiments.

In the 1970s Russell Targ and Harold Puthoff carried out some of the best known tests on thought and image transference. They wished to ascertain the reality of telepathic transmission between a 'sender' and a 'receiver.' In an experiment that has since been repeatedly performed, the receiver is placed in a sealed, opaque and electrically shielded chamber, and the sender in another room where he or she is subjected to bright flashes of light at regular intervals. Electroencephalograph (EEG) machines are employed to register the brain-wave patterns of both sender and receiver. As expected, the sender exhibits the rhythmic brain waves that normally accompany exposure to bright flashes of light. But, in a significant number of cases, following a brief interval also the receiver begins to produce the same patterns, although he or she is not exposed to the flashes and is not receiving sensory clues from the sender.

Targ and Puthoff also designed so-called remote viewing experiments. (Targ recently defined remote viewing as the acquisition and description, by mental means, of verifiable information about the physical universe blocked from ordinary perception by distance or shielding and generally considered to be secure from such access.[3]) In these experiments sender and receiver are separated by distances that preclude any form of sensory communication between them. At a site chosen at random, the sender acts as a 'beacon' while the receiver tries to pick up what he or she is seeing. To document the receiver's impressions verbal descriptions are given, accompanied at times by sketches. In the original studies, judges found that the descriptions of the sketches matched on the average 66 percent of the time the characteristics of the site actually seen by the beacon.[4] A 1979 follow-up experiment conducted by Russell Targ incorporated all revisions in methodology suggested by the critics of earlier studies. It involved six inexperienced volunteers each of whom attempted to describe six randomly selected distant locations visited by the experimenters. Four of the subjects achieved independent statistical significance in their six trials; overall, they achieved more than fifty percent first-place matches.[5]

In the period of ten years since Targ and Puthoff first published their Stanford experiments, twenty-four attempted replications have been carried out, with more than half of them reported as successful.[6] Distances ranged from half a mile to several thousand miles, and the success rate, regardless of where they were carried out, and by whom, was generally around fifty percent. The most successful viewers proved to be those who were relaxed, attentive, or meditative. Their preliminary impression was often described as a gentle and fleeting form which gradually evolves into an integrated image.

Images can also be transmitted while the receiver is asleep. Over several decades, Stanley Krippner and his associates carried out 'dream ESP experiments' at the Dream Laboratory of Maimondes Hospital in New York. The experiments followed a simple yet effective protocol. The volunteer, who spends the night at the laboratory, has electrodes attached to his or her head to monitor brain waves and eye movements; there is no sensory contact with the sender until the next morning. One of the experimenters throws dice that, in combination with a random number table, gives a number that corresponds to a sealed envelope containing an art print. The envelope is opened when the sender reaches a private room in a distant part of the hospital; thereafter the sender spends the night concentrating on the print.

The experimenters wake the receiver by intercom when the monitor indicates the end of a period of rapid eye-movement (REM) sleep. The subject is then asked to describe any dream he or she might had before awakening. The comments are re-corded, together with the contents of an interview next morning when the subject is asked to associate with the remembered dream contents. The interview is conducted 'double blind' — neither the subject nor the experimenters know which art print had been selected the night before.

Using data taken from the first night each volunteer spent at the dream laboratory, the series of experiments that took place between 1964 and 1969 produced 62 nights of data for analysis. The result was a significant correlation between the art print selected for a given night and the recipient's dreams on that night. Dividing the results into four categories, ranging from 'high hit' to 'low miss,' there were a total of 18 high hits, 29 low hits, seven high misses, and eight low misses.[7]

A further variety of ESP involves the transmission of physiological changes triggered in the receiver (usually another person, but on occasion a microorganism, plant, or animal) by the mental processes of the sender. Traditionally, such 'bio-psychokinetic' effects were produced by specially gifted healers, who would 'send' what they claimed to be subtle forms of energy to their patients. The negative variety of effects came under the heading of woodoo or black magic; they were—and to some extent still are—common in the practice of shamans and witch doctors.

Being largely anecdotal, bio-PK was of interest mainly to anthropologists; it was dismissed by the medical community. Lately, however, healing and other physiological effects have been investigated in controlled experiments, where either a sufficient number of trials or a sufficient number of test subjects permitted a quantitative evaluation. William Braud and Marilyn Schlitz carried out hundreds of trials with rigorous controls regarding the impact of the mental imagery of 'senders' on the physiology of 'receivers'. The latter were both distant, and unaware that such imagery was directed to them. Braud and Schlitz claim to have established that the mental images of a person can 'reach out' over space and cause changes in the physiology of a distant person—effects that are comparable to those one's own mental processes produce in one's body.[8]

Experiments have also made use of intercessory prayer to achieve effects on the physiology of test persons. Randolph Byrd made a ten-month computer-assisted study of the medical histories of patients admitted to the coronary care unit at San Francisco General Hospital. He formed a group of experimenters made up of ordinary people whose only common characteristic was a habit of regular prayer in Catholic or Protestant congregations around the country. The selected people were asked to pray for the recovery of a group of 192 patients; another set of 210 patients, for whom nobody prayed in the experiment, made up the control group. Rigid criteria were used: the selection was randomized and the experiment was carried out double blind, with neither the patients, nor the nurses and doctors knowing which patients belonged to which group. The experimenters were given the names of the patients, some information about their heart condition, and were asked to pray for them every day. Since each experimenter could pray for several patients, each patient had between five and seven people praying for him or her. The results were statistically significant. The prayed-for group was five times less likely than the control group to require antibiotics (three compared to sixteen patients); it was three times less likely to develop pulmonary edema (six versus eighteen patients); none in the prayed-for group required endotracheal incubation (while twelve patients in the control group did); and fewer patients died in the former than in the latter group (though this particular result was statistically not significant). It did not matter how close or far the patients were to those who prayed for them, nor did the manner of praying make any difference. Only the fact of concentrated and repeated prayer was a factor, without regard to whom the prayer was addressed and where the prayers took place.[9]

Byrd's experiment with 'intercessory prayer' has been followed by numerous others, exploring a variety of bio-PK and ESP phenomena. Several of the experiments have been statistically analyzed, and sets of so analyzed experiments were then subjected to meta-analysis, where data from a number of experiments were gathered and analyzed together.[10] Paranormal occurrences have been found in the majority of the analyzed cases. Daniel Benor, who reviewed experiments on the effects of intercessory prayer on healing, found that of 131 experiments of which statistical results have been published up to 1993, 56 had a probability value (p) of less than .01, and a further 21 had probability values between .02 and .05. He also found that of 155 controlled studies on psychic healing (involving test objects as varied as enzymes, yeasts, bacteria, red and white blood cells, cancer cells, plants, mice, as well as human subjects) 67, or 43 percent, have had probability values of less than .01, and another 23 (15 percent) values in the range of .02 to .05.[11]

Similar results have been obtained in other meta-analyses. R. Rosenthal analyzed ESP experiments using the Ganzfeld technique (where variety in the sensory input to the subject is drastically reduced in the attempt to elicit paranormal perceptions) and found that of 28 statistically evaluated experiments 23 (or 82 percent) had effect-sizes greater than zero. The median effect-size was .32 and the mean .28, where the purely random processes of the known mechanisms would have given .25.[12] Honorton *et al.* carried out similar meta-analyses on eleven series of Ganzfeld experiments involving eight experimenters, 240 test subjects, and 354 trials. They found an average effect size of .34 (compared likewise with .25 as given by pure chance).[13]

Results such as these demand to be taken seriously, the more so as both physical and psychological variables affect the scores in statistically significant and predictable ways. Stanley Krippner and Michael Persinger noted that dream telepathy scores were significantly lower in nights when there were electrical storms and a high level of geomagnetic activity in comparison with other nights.[14] Mario Varvoglis, in turn, observed that in general extroverts score higher than introverts in ESP and bio-PK experiments, the same as the creative and artistically gifted compared with the non-creative, and the internally oriented compared with the externally oriented. Also those with low defensive attitudes and higher levels of belief and confidence achieve higher scores than subjects with high defensive orientation and more scepticism.[15]

Anomalous recall

Some phenomena of long-term recall likewise suggest inputs and interactions beyond the range of ordinary sensory communication. Since Elisabeth Kübler-Ross' classic studies, near-death experiences (NDEs) have been systematically investigated by clinical psychologists and other specialized researchers. Raymond Moody Jr. concluded that it is now 'clearly established' that the experience of a significant proportion of the people who are revived following close calls with death is quite similar from case to case, regardless of the patient's age, sex, religious, cultural, educational or socioeconomic background.[16] The experience itself is more widespread than is generally recognized; a survey conducted by George Gallup, Jr. in 1982 found that some eight million adults in the US alone have undergone them.[17] The near-death experience alters the subsequent course of people's lives: they are no longer fearful of death but focus on the importance of the present with enhanced love and concern for others.

Memories seemingly of one's entire life form an important element of the near-death experience: thirty-two percent of the eight million people reported in the Gallup poll said that 'life-reviews' were a part of their near-death experience. David Lorimer distinguished two kinds of recall: panoramic memory, and the life-review itself. Panoramic memory consists of a display of images and memories with little or no direct emotional involvement on the part of the subject; while life-review, although superficially similar, manifests emotional involvement as well as moral assessment. The clarity of mental processes is noteworthy in both; it is especially vivid in panoramic memory, where there is a remarkable speed, reality and accuracy in the images that flash across the mind. The time-sequence of the memories may vary: some start in early childhood and move towards the present; others start in the present and move backwards to childhood; still others come superposed, as if in a holographic clump. To the subjects it appears that everything they have ever experienced in their lifetime is being recalled; no thought, no incident, appears to have been lost.[18]

John von Neumann calculated that an individual accumulates some $2.8 \cdot 10^{20}$ bits of information during his or her lifetime. Thus when NDEs suggest that the quasi-totality of a person's lifetime experience is subject to recall, a truly staggering quantity of data is involved. There is evidence, however, that on occasion the human mind has access to a memory bank that is even vaster. Psychotherapists who 'regress' patients to early childhood find that they can go further back in time, uncovering memories of the womb and of birth. Frequently the flow of memories reaches to putative prior lifetimes.

Therapists engage in 'regression therapy' not for the intrinsic interest of this esoteric memory flow, but because the images and events recalled by their patients often relieve their traumas and neuroses. They find that many patients seem able to recall several past lives, covering a vast time span. According to Thorwald Detlefsen, the series of 're-incarnations' runs into the hundreds and may encompass 12,000 years. Stanislav Grof claims to have hypno-regressed subjects even to the state of animal ancestors. Patients of all ages tell stories of prior life-experiences, often associated with present problems and neuroses. Detlefsen's case histories include the story of a patient who could not see in an otherwise functional eye; he came up with the memory of being a medieval soldier whose eye was pierced by an arrow. A patient of Morris Netherton, suffering from ulcerative colitis, re-lived the sensations of an eight-year-old girl shot at a mass grave by Nazi soldiers, while Roger Woolger's patient, who complained of rigid neck and shoulders, recalled committing suicide by hanging as a Dutch painter.[19]

It is not clear whether such images and experiences are products of the subjects' imagination, or come to them paranormally from an external source. On occasion evidence has been uncovered that subjects had prior information about the recalled persons, times, or places. But in other cases the information produced by the subjects contained elements that are not likely to have been previously available to them, such as obscure (but subsequently verified) historical and geographical particulars. The cases most difficult to account for are those in which children recount experiences that seem to stem from previous lives—and when the recalled experiences refer to the lives of actual personages.[20]

Another difficult case is xenoglossy, the speaking of a language of which the subject had no previous knowledge. The assumption of chance acquaintance with some elements of that language will not do: in several recorded cases regressed subjects engaged in prolonged and fluent conversations in the given language.[21]

Though findings such as these are impressive, they do not disclose the nature of the processes that would be responsible for them. Reviewing demonstration research and meta-analysis in parapsychology, a reputed group of investigators noted that research does not give information about how or why the observed events have occurred. Meta-analysis, in the words of another investigator, is 'glorified literature review'. Even at its best, it cannot determine what is *true*, but can only attempt to measure what *is*.[22] In a similar vein, Russell Targ noted that in the last analysis 'all is data'. Psi-research's significant accomplishment over the past two decades is that ESP and other varieties of psi can be demonstrated when needed for study and investigation: it is no longer elusive. The task is now to discover where it comes from.[23]

The interpretation of interactive experience

Let us consider the task proposed by Targ. Current theories of perception and related brain function indicates that perceptual processes in the brain are creative rather than passive: they create the contents that appear in consciousness on the basis of analyzing distributed wavepatterns in the ambient fields. The brain's signal-analyzing capacity is extremely sophisticated: its range may extend all the way to the quantum level. It is reasonable, then, to inquire whether the wave-patterns that create the contents of consciousness in the brain may include interference patterns in the vacuum's holofield.*

There is evidence that the brain may be capable of accessing and processing signals of vacuum-fluctuation magnitude. First of all, the receptive regions of the brain are known to be permanently in a state characterized by chaos, and in that state they are ultrasensitive to the most subtle alterations in the ambient fields. Vast collections of neurons shift abruptly

---

* We should specify that it is unlikely that the signals accessed by the brain would concern the totality of the information encoded in the local topology of the vacuum (that magnitude of information-load would be literally mind-boggling), nor is it plausible that the accessed set of signals would be limited to the brain's own neural processes. The brain is likely to be 'in-formed' by its own past processes as well as by a specific selection of signals originating in the electromagnetic (and other) fields. The latter may comprise the scalar-mediated holofield of the vacuum: the hypothesis that we shall now proceed to investigate.

and simultaneously from one complex activity pattern to another in response to extremely fine variations. Within the ten billion neurons of the brain, each with an average of twenty thousand interconnections, the action potential of the smallest neuronal cluster creates a 'butterfly effect' that triggers massive gravitation towards one or another of the chaotic attractors. These attractors may amplify even vacuum-level fluctuations so as to produce effects on the brain's information processing.

Secondly, evidence is forthcoming that scalar waves may be biologically even more active than their electromagnetic counterparts. Though the interaction of scalar and electromagnetic waves is not known in detail, Glen Rein proposed that scalar energy is transduced into linear electromagnetic energy in the body by liquid crystals in the cell membrane and solid crystals in the blood and various biological tissues.[24] His experiments on nerve cells in tissue culture indicate that scalar energy can modulate basic biochemical communication between nerve cells as mediated by the known neurotransmitters.[25] Rein used a particular type of nerve cell, *PC12*, originally isolated from a rat adrenal phaeochromocytoma, because its neurotransmitter properties have been studied and were shown to be similar to those occurring during normal synaptic transmission in the brain. He investigated the noradrenaline uptake of the cells, using the same protocol that was used when determining the effects of EM fields on this uptake. Overall, the uptake of noradrenaline in cells exposed to an 8 Hz scalar wave field was inhibited by 19.5 percent compared to cells in the control dish. Rein concluded that scalar energy has a direct effect on nerve cells, and possibly on other cells in the body.

There is further evidence pertinent to the organism's—and hence the brain's—sensitivity to vacuum-level inputs. Holographic functions, whether in the brain or in artificial systems, require coherent nonlinear interaction between neuronal networks and/or pre- and post-synaptic neurons. In biological systems coherent interactions have been noted within molecules, between molecules, as well as among dipole clusters in distinct cellular and anatomical structures. In the past, such phenomena have been explained in terms of long-range electromagnetic correlations between physically separated oscillating electric dipoles. Recently, however, an alternative explanation has surfaced. As indicated in Chapter 3, the new concept makes reference to the Josephson effect. According to quantum field theory, Josephson junctions generate fields of quantum potentials (consisting of a magnetic vector potential and an electrostatic scalar potential) which in turn modulate the connection between the correlated superconductors or cellular systems.

This indicates that fields of quantum potentials may mediate communication between physically separate assemblies of neurons in the brain, constituting an underlying regulatory system that alters non-synaptic communication between assemblies of neurons.[26] An explanation of this kind is required to account for the quasi-instantaneous synchronization of large assemblies of neurons in diverse parts of the brain. Such synchronization, which experiments with animals have shown to occur at 40 Hz, takes place within 1/1,000 to 1/10,000 sec. In a ten-centimeter diameter brain, this would require a transmission velocity of 100 to 1,000 m/sec. This, however, vastly exceeds the known transmission speed of axonic and dendritic connections between neurons.

Fields of quantum potentials can be interpreted as electromagnetic fields mediated by scalar wave-interference patterns. Scalar patterns that are isomorphic with the Fourier transform of particular neural networks could mediate electromagnetic waves in the brain, and the thus mediated electromagnetic waves would then interact with electrons and neurons in the cerebral networks. The process may unfold on the basis of resonance: the pertinent scalar frequencies coincide with the electromagnetic wave frequencies observed during cell division (see Section [i] in Part Three). A review of some of the interactive mind-phenomena described above can shed light on this proposition.

Near-death and past-life experiences revisited

Recall at the portals of death, it is sometimes said, is merely a by-product of progressively dissolving synapses and decaying brain tissue. This, however, conflicts with the noted sharpness and speed of the recalled images. The NDE memory phenomenon points to the availability of long-term recall in humans, though such recall is not well understood in neurophysiology.

Short-term memory can be relatively well understood in reference to the formation and reformation of neuronal networks in the cortex, but long-term seems to call for some variety of traces or 'engrams' that would modify the synapses between neurons. As Sir John Eccles remarked, "we have to suppose that long-term memories are somehow encoded in the neuronal connectivities of the brain. We are thus led to conjecture that the structural basis of memory lies in the enduring modification of the synapses."[27]

However, the search for engrams, or other enduring synaptic modifications through which experiences would be permanently stored, has proved fruitless. The search began in a systematic fashion in the 1940s with Karl Lashley's celebrated series of animal experiments. Lashley was trying to find permanent engrams in the brain of rats by the expedient of teaching them specific behavioral routines and then cutting out various parts of their cortex to see where the instructions for the routines would be stored. He cut out larger and larger segments of brain tissue, but found no correlation between brain area and recall of the routine: the test animals' recall degenerated proportionately to the amount of tissue removed, but never ceased entirely.[28] The puzzle as to where the traces of the routines are stored, has not been solved to this day. J.Z. Young conceded that, even if most neuroscientists believe in a theory of synaptic change, there is little direct evidence of the details of it.[29]

Edelman's theory of neural group selection, while it provides a convincing account of the selective modification of fixed behavioral patterns (including the cognitive correlates of the modified behaviors), is less convincing when it comes to accounting for long-term memory. Cognitive functions are explained in terms of structurally distinct neuronal groups that range anywhere from one hundred to one million cells; such groups are said to respond as a unit to a signal conveyed to them. Each group responds to a specific subset of signal types; these are the subsets that generate attention-responses in mental processing. Since the signals select particular neuronal groups, the groups are in competition with each other in regard to their activation. For this reason the theory of neuronal group selection is known as 'neural Darwinism.'

The three tenets of TNGS relate to the development of the anatomy of the brain; the selection from this anatomy during experience; and the emergence of behaviorally important functions through a process of signalling between the maps that result in the brain. The primary processes of development lead to the formation of the neuroanatomy characteristic of a given species. The developmental process is selectional, involving populations of neurons engaged in topobiological competition. A population of variant groups of neurons in a given brain region, comprising neural networks, constitutes what Edelman calls the *primary repertoire*. The genetic code, rather than providing specific instructions for this repertoire, imposes a set of constraints on the process of its selection.

A further mechanism of selection does not generally involve an alteration of anatomical patterns: it is based on a selective strengthening or weakening of synaptic connections during behavior. This selection process produces a set of variant functional circuits called the *secondary repertoire*. The primary and secondary repertoires form maps in the brain, connected by massively parallel and reciprocal connections. Correlation and coordination of the selection events are achieved by re-entrant signaling and the strengthening of interconnections between the maps within a given segment of time. Thus mental development involves the selection of pre-existing neuronal groups by incoming signals, and the amalgamation of the groups into higher order configurations.

Memory for the TNGS is essentially the ability to repeat a performance. Alterations in the synaptic strength of neuronal groups in a global mapping provide the biochemical basis of memory; the phenomenon is a population property created in the context of continual dynamic changes in the synaptic populations. Evidently, in such a system memory cannot issue in a stereotypic form of recall: recall must change under the influence of the continually changing context. Memory, after all, wrote Edelman, is the result of a process of continual recategorization, as perceptual categories are altered by the ongoing behavior of the animal.[30]

It is not accidental that Edelman cited animals in connection with his account of memory: the TNGS offers a good fit in regard to some varieties of animal memory. In animals what ethologists sometimes call 'genetic memory' constitutes the primary repertoire. However, as in higher animals this is not sufficient to ensure survival, the rigidities of the behavioral routines scripted by this repertoire are supplemented by learned behaviors. The latter may well obtain due to re-transcriptions in the neural networks of the animal brain, i.e., through the formation of the secondary repertoire. Birds, for example, such as tits hunt insects randomly provided that various species abound in their milieu (this behavior being instructed by genetic memory), but if one insect species is present in larger numbers than the others, they begin to hunt that species preferentially, using the neural memory of the secondary repertoire. As Beritashvili's experiments show, even fish 'recall' the location of the box where they were fed, though such memory lasts less than ten seconds. The corresponding memory in frogs and turtles comes to several minutes, in dogs to several hours, and in baboons to about six weeks.

It is questionable, however, whether memory in humans is restricted to a temporary modification of performance through synaptic retranscription. Edelman, who recognizes that humans have developed a much richer set of psychological functions than animals, admitted that higher functions alter the meaning of what it is to have memory, but claimed nevertheless that no new principles beyond selection and re-entry are necessary to gain new memory functions: new orderings of connections in the brain are all that is needed.[31] This claim does not appear well founded. While some forms of memory in humans could result from continual dynamic changes in synaptic populations within global mappings—for example, short-term behavior-modifying memory—human memory is not limited to these forms: it also includes the vivid and often surprisingly accurate recall of a complex sequence of events with a vast series of associated images. These events and images could have been experienced many years ago; and they need not have any immediate behavioral correlates.

The type of memory for which TNGS, as other synaptic-change-based theories, fails is that in which events and images experienced in the distant past reappear vividly, accurately, and in detail. In a completely different vein, David Lorimer notes that the only picture within which total 'life-review' experiences make sense is one of an "interconnected web of creation, a holographic mesh in which the parts are related to the Whole and through the Whole to each other by empathetic resonance." This must be the sort of Whole, Lorimer added, in which we and the rest of creation have our being; a consciousness-field in which we are interdependent strands.[32] His explanation, while on the right tack, is unnecessarily impressionistic; it needs to be—and can be—improved upon.

In this context it is useful to recall Lashley's own conclusions. Faced with unexpectedly distributed memory in rats, he speculated that, without regard to particular nerve cells, animal behavior must be determined by 'masses of excitation' within general fields of activity. He likened these fields to the force fields that determine form during embryogenesis: similar lines of forces could create patterns in cortical tissue.[33] This may offer a fruitful approach, yet not many neuroscientists have followed it up.

The variety of field theory espoused here is an alternative both to re-transcription-based neural network theories and to impressionistic insights. Its principal innovative feature is that it claims that the events, images, and other items recalled in long-term memory are not stored *in* the brain; they are only *accessed* by the brain from an ambient field. In this concept long-term memory is not physically located in the cerebral networks; the latter act only as the transducer of signals received from the vacuum holofield, which then functions as an extrasomatic memory store. Retrieval from that field occurs on the holographic principle. Interference patterns are re-converted into the image of the objects or events that have originally recorded the given pattern, following Fourier transformation rules.

This concept shows a good fit with the observed features of memory recall. First, recall is not localized in brain tissue but appears distributed over wide brain regions that, following Pribram and likeminded brain theorists, are holographic receptor patches.

Second, recall has an associative property: whenever any fragment of a recorded information is presented to attention, that fragment acts as a memory address for the recovery of a wide range of associated information—consistently with the fact that in a hologram any small part includes the full set of recorded information. Third, recall consists of a complex set of data (visual, acoustic and related memories), often in the form of time-varying sets of data ('moving-scene memories'), suggesting mechanisms similar to those of multiplex holography. Fourth, access time in the brain is not related to the scanning time of the stored experiences but is dependent mainly on the level of attention of the subject and the emotional intensity accompanying the recall—thus there is no evidence that mnemonic traces in the brain would be stored in organized files, tree-shaped archives or other mechanisms similar to information storage in libraries and conventional computers.

The above factors suggest that recall is likely to be a holographic process occurring in an interaction between the brain of the subject and an ambient holographic field. This allows a reinterpretation of long-term and transpersonal varieties of recall. Conservative observers dismiss these phenomena as fantasy or illusion, while speculative investigators speak of reincarnation. It may well be, however, that reality warrants neither the one nor the other assumption.

Transpersonal memory can be seen as the expansion of the 'band-width' of the vacuum-fluctuation receptivity of the brain's holographic receptor patches. Pribram has shown that holonomic information processing in the brain requires Gabor-transforms that limit the otherwise infinite Fourier-transforms so as to produce a precise match between the holographic patch in the receptive regions and the incoming waveforms. In sensory perception a Gaussian envelope constrains the Fourier-transforms and yields the patchwork Gabor-transforms. In transpersonal experience the Gaussian constrains appear to be relaxed—a condition that yields the spaceless and timeless oceanic sensation in which the nervous system becomes, in Pribram's words, "attuned to the holographic aspects of— the holograph-like order in—the universe."[34] (Evidently, some level of relaxation of the Guassian constraints is necessary even if ordinary forms of recall are to function. The brain ages, like the rest of the body, and its typical configuration of neuronal structures undergoes subtle alterations. High selectivity in the Gabor-transforms would restrict the re-translation of field-coded experiences to a narrowly limited timespan.)

If the possibility of transpersonal recall is to be accounted for, we must assume that the brain's operative transforms have a non-negligible band-width. They are operative within a *range* of multidimensional waveforms, rather than within highly specific frequencies. It follows that in states of abnormal functioning (altered states, and immature or degenerative brain states), the brain may be unable to distinguish between waveforms within a range of adjacent frequencies. When two wave interference-patterns fall within the tolerance-range of the transforms, altered-state or immature brains decode both of them equally. This is the case even if a pattern codes the $3n$-dimensional configuration-space of another person's cerebral networks, rather than the subject's own. In that event the subject recalls that person's experiences as if they were his or her own.

Altered states of consciousness (ASCs), often seen as a prerequisite for the occurrence of transpersonal experience, appear to enlarge the band-width of Gabor-transforms in the subject's brain, so that a larger variety of scalar waveforms falls within their operative range. This is what seems to take place in regression therapy. Here a suitably altered state is induced in the patient by the therapist, and in this condition the patient fails to distinguish between the recall of his or her own experiences and the related experiences of one or more other persons. Perceiving the images and events as if they were records of his or her own experiences, the patient comes up with accounts of previous lives.

The relatively frequent occurrence of pastlife experiences in young children has a related but somewhat different explanation: here the enlargement of the brain's operative band-width would be due to the immaturity of the cerebral networks. EEG records show that the brains of children function enduringly in the alpha-wave mode that in adults occurs mainly in ASCs—until the age of five or six years, the beta waves of normal waking consciousness are seldom present. Thus the child's brain may not be able to dependably discriminate the waveform inputs that convey the experiences of a brief lifetime from sufficiently isomorphic inputs that convey the experiences of other persons. (Children, we should add, are also not constrained by the anomaly-repressive mechanisms typical of adults, so that they could all the more readily become aware of images and impressions that are not traces of their own experiences, but traces of the experiences of people whose brain states happen to match theirs).

The match between a child's own memory recall and the recall of another person's isomorphically coded memories may be triggered or reinforced by physiological abnormalities. Stevenson noted cases of birthmarks, birth defects, and other bodily deformations that match misfortunes that befell persons whom the deformed or marked child 'remembers.'[35] On first sight, these cases suggest that the deformed child has somehow acquired the bodily traits of the person he or she remembers—i.e., that the child 'reincarnates' that person. However, the QVI hypothesis offers a less speculative explanation. The causal relationship between mental and physiological events may be the reverse of that assumed in the reincarnationist thesis. It is not that an individual would receive (at birth or perhaps already at inception) the soul or personality of a deceased person, and that this mental factor would then produce his or her physiological and psychological characteristics. Rather, the physiological and psychological traits occur first, due to independent causes. If and when they occur, they would enable the child's brain to operate the transforms that match the corresponding experiences. Thus when a Turkish boy with a severely malformed ear 'remembers' the life of a man who was fatally wounded by a shotgun discharged at close range, and when an Indian boy, born with absent fingers on one hand, recalls the life of another boy in a neighboring village who had put his hand into the blades of a fodder-chopping machine, it is not that they would reincarnate the soul or personality of the individuals whose memories surface in their consciousness. It is more likely that their brain picks up the images and events from the ambient holofield that match their bodily deformations.

A fresh look at E S P

The transference of thoughts or images between persons who are not in sense-mediated contact with one another finds explanation in basically similar terms. As we have seen, a close matching of the relevant brain states seems to be a precondition of effective transference. The extrasensory transference of thoughts, images, emotions, intuitions, and even of physical sensations, occurs most readily when sender and receiver are genetically related (truly striking instances are furnished by identical twins); when they are bound by close emotional ties (such as mothers and sons, spouses or lovers); and when they are in an altered state of consciousness. It appears that matching states can be 'hard-wired' by genetic predisposition—as in identical twins—or produced by emotional ties and empathy, especially in conditions of crisis and trauma. Altered states relax the Gaussian constraints on the brain's receptive patches, enlarge the band-width of the operative transforms and thereby extend the range of input receptivity.

The ability of altered states to convey information in the extrasensory mode is supported by data coming to light in EEG experiments. Experiments carried out in Italy with the 'brain holo-tester' (a device designed to measure levels of synchronization in EEG patterns between the left and right hemisphere of one person, as well as hemispheric synchronization between different persons) indicate that in the altered state of deep meditation the synchronization of the left and right hemispheres increases dramatically. Moreover when two persons meditate together, their respective EEG patterns become highly synchronized, even though no sensory signal is passing between them.[36]

The physiological effects of spontaneous communication with another person are highlighted in bio-PK (biological psychokinesis). Already in 1967, Dean and Nash carried out experiments in which a sender attempted to send words to a receiver of which some were emotionally charged and others not. The subjects had no knowledge when the corresponding transmissions would occur—they were timed randomly, not under the control of either senders and experimenters. The latter, however, controlled blood flow in the receivers with a plethysmograph. They found significant physiological effects coinciding with the sending of emotionally charged words.[37] Subsequently, at Maimonides Hospital in New York, Kelly, Varvoglis and Keane had senders visualize short but intense video documentaries on a screen and tested receivers on a distant location for galvanic skin responses. The physiological effects on the receivers coincided with the times when the senders viewed the documentaries.[38]

It is noteworthy that in psi experiments physiological responses tend to occur more frequently and reliably than conscious responses. In a study carried out by Deborah Delanoy and Sunita Sah, senders were asked to project positive, happy memories as well as emotionally neutral thoughts to percipients in another part of the building. Thirty-two sender-percipient pairs each participated in one session of sixty-four 30-second periods, during which 16 positive and 16 neutral sending intervals were interspersed with 32 rest-periods. Electrodermal activity (EDA) was measured for all trials while conscious responses were obtained for the first or the second half of each session. The subjects showed significantly greater EDA activation during positive emotional periods than during neutral periods ($p = 0.043$), while the conscious responses were not above random probability.[39] Similar patterns have been found by other investigators, notably Tart, and Targ and Puthoff.[40]

Bio-PK effects are not illusory, but they are also not necessarily what they appear to be: namely, the effects of the mind of one person on the body of another. A survey of natural healing experiments by S.A. Schouten showed little evidence that 'the effect of the method' would be responsible for the observed healing. Healing effects due to the effectiveness of the healer's influence on the body of the patient were found to be either weak or non-existent, while psychological variables associated with the patient, and with the healer-patient interaction, were seen as the factors that contributed most to the healing process. The positive effects were achieved by what Schouten termed 'psycho-logical factors associated with patients and healers.'[41]

Schouten claimed that the concept of the paranormal cannot really provide an explanation for psychic healing. This, however, may be unduly sceptical. We must entertain the possibility that on occasion healing may occur by a transference of a state of mind from the healer to the patient. This conclusion is also reached by Varvoglis. It is possible, he observed, that the periods when physiological effects are noted are those during which the sender transmits a telepathic message to the receiver, prompting excitation or relaxation. The effects could then be produced by the receiver psychosomatically.[42]

'Tele-psychosomatic' (rather than directly telesomatic) transference calls for an informational link between the brain of the healer and the brain of the patient. In regard to the healer-patient interaction, the explanation outlined above in regard to telepathy can convey a basic understanding. It suggests that bio-PK is not the effect of an intentional mind on a biological system. Psi experimenters themselves question this: they are not agreed on the interpretation of the results, not even on whether psychokinesis (the transmission of some kind of force or energy from an intentional mind to a receiving organism) actually takes place. Varvoglis notes that most statistical PK experiments involving mobile or sensitive targets can be accounted for through precognition, clairvoyance, or in general as a form of intuitive data sorting.[43] Stanford, in turn, questions whether there is any convincing evidence for a PK 'force' monitored and guided by sensory or extrasensory means. His 'conformance-behavior model' maintains that PK is a form of direct causation, independent of both energetic factors and factors related to information processing; it occurs only to the extent that it is relevant to particular needs or dispositions.[44] Others, such as Millar, agree that PK is not a function of energetic or spatiotemporal factors and suggest that the outcome of PK experiments is determined by the act of observing the outcomes. Sensory feedback 'triggers' an imposition of information on the observed event, even if the event has been generated before the observation.[45] And in the view maintained by Walker, the will of the observer collapses the wave-function of the target system and creates the noted PK-effect.[46]

Evidently, much more work is required before the nature of bio-PK can be convincingly clarified. This, however, does not cast doubt on the occurrence of the corresponding psi-phenomena, nor alter the cogency of the postulate that such phenomena indicate spontaneous interaction between the brain/minds of individuals—interactions that obtain even when the subjects are separated in space and time.

References

1. Ervin Laszlo, *Introduction to Systems Philosophy: Toward a New Paradigm of Contemporary Thought.* Gordon and Breach, New York 1972, reprinted 1987.

2. A.P. Elkin, *The Australian Aborigines,* Angus & Robertson, Sydney 1942.

3. Russell Targ, 'Remote viewing replication: evaluated by concept analysis.' Dick J. Bierman (ed.), *Proceedings of Presented Papers,* Parapsychological Association 37th Annual Convention, University of Amsterdam, 1994.

4. Russell Targ and Harold Puthoff, 'Information transmission under conditions of sensory shielding,' *Nature,* 252, 5476 (1974)

5. Russell Targ, 'Remote viewing...' *op. cit.*

6. G.M. Hansen, M. Schlitz and C. Tart, 'Summary of remote viewing research,' in Russell Targ and K. Harary, *The Mind Race. 1972-1982.* Villard, New York 1984.

7. Michael A. Persinger and Stanley Krippner, 'Dream ESP experiments and geomagnetic activity,' *The Journal of the American Society for Psychical Research,* **83** (1989);
   M. Ullman and S. Krippner, *Dream Studies and Telepathy: An Experimental Approach,* Parapsychology Foundation, New York 1970.

8. W. Braud and M. Schlitz, 'Psychokinetic influence on electrodermal activity. *Journal of Parapsychology,* **47** (1983).

9. R.C. Byrd, 'Positive therapeutic effects of intercessory prayer in a coronary care population,' *Southern Medical Journal,* **81**,7 (1988).

10. S. Krippner, W. Braud, I.L. Child, J. Palmer, K.R. Rao, M. Schlitz, R.A.White, and J. Utts, 'Demonstration research and meta-analysis in parapsychology.' *Journal of Parapsychology,* **57** (1993).

11. Daniel J. Benor, *Healing Research: Holistic Energy Medicine and Spiritual Healing,* Helix Verlag, Munich, 1993.

12. R. Rosenthal, 'Combining results of independent studies,' *Psychological Bulletin,* **85** (1978).

13. C. Honorton, R. Berger, M. Varvoglis, M. Quant., P. Derr, E. Schechter, and D. Ferrari, 'Psi-communication in the Ganzfeld: Experiments with an automated testing system and a comparison with a meta-analysis of earlier studies.' *Journal of Parapsychology,* **54** (1990).

14. M.A. Persinger and S. Krippner, 'Dream ESP experiments and geomagnetic activity. *Journal of the American Society for Psychical Research,* **83** (1989), 101-116.

15. Mario Varvoglis, 'Nonlocality on a human scale: Psi and consciousness research.' LRIP, Paris 1994 (mimeo).

16. Raymond Moody, Jr., *Life After Life,* Mockingbird Books, Covington 1975; —, Foreword, in David Lorimer, *Whole In One: The Near-Death Experience and the Ethic of Interconnectedness,* Arkana, London 1990.

17. Cited by Raymond A. Moody, Jr. in *The Light Beyond,* Bantam Books, New York 1988.

18. Lorimer, *Whole In One, op.cit.,* Ch.1.

19. Roger Woolger, *Other Lives, Other Selves,* Doubleday, New York 1987; Thorwald Detlefsen, *Schicksal als Chance* [Fate As Opportunity], Bertelsmann Verlag, Munich 1979; Morris Netherton and Nancy Shiffrin, *Past Lives Therapy,* William Morrow, New York 1978.

20. Ian Stevenson, *Children Who Remember Previous Lives,* University Press of Virginia, Charlottesville 1987.

21. —, *Unlearned Lanquage: New Studies in Xenoglossy,* University Press of Virginia, Charlottesville 1984.

22. T. Adler, 'Meta-analysis offers precision estimates. ' *APA Monitor,* September 1990, p.4.

23. Russell Targ, 'What I see when I close my eyes,' *Journal of Scientific Exploration,* **8**,1 (1994), p. 117.

24. Glen Rein, 'Mechanisms of psychic perception,' in J. Millay, and S-P. Sirag (eds.), [in press]

25. —, Biological interactions with scalar energy-cellular mechanisms of action, *Proceedings of the 7th International Association of Psychotronics Research,* Georgia, December 1988.

26. —, 'Modulation of neurotransmitter function by quantum fields.' *Planetary Association for Clean Energ,y* **6**, 4 (1993).

27. John Eccles and Daniel N. Robinson, *The Wonder of Being Human*, Shambhala Publications, London 1985.

28. Karl Lashley, 'The problem of cerebral organization in vision,' Biological Symposia, **VII**, *Visual Mechanisms,* Jacques Cattell Press, Lancaster 1942.

29. J. Z. Young, 'Memory,' *Oxford Companion to the Mind*, Richard Gregory (ed.). Oxford University Press, UK, 1987.

30. Gerald M. Edelman, *Bright Air, Brilliant Fire On the Matter of Mind*, New York, Basic Books, 1992.*op.cit.*, p.102.

31. *Ibid*, 107-8.

32. David Lorimer, *Whole in One, op.cit.*, 22.

33. Karl Lashley, 'The problem of cerebral organization in vision,' *op. cit.*

34. Karl Pribram, *Brain and Perception: Holonomy and Structure in Figural Processing*, The MacEachran Lectures. Lawrence Erlbaum, Hillsdale, NJ, 1991, Chapter 9.

35. Ian Stevenson, 'Birthmarks and birthdefects corresponding to wounds on deceased persons,' *Journal of Scientific Exploration*, **7**,4 (1993).

36. *Cyber*, Milan, 40, November 1992.

37. E.D. Dean and C.B. Nash, 'Coincident plethysmograph results under controlled conditions,' *Journal of the Society of Psychical Research*, **44** (1967).

38. M.T. Kelly, M.P. Varvoglis and P. Keane, 'Physiological response during psi and sensory presentation of an arousing stimulus,' *Research in Parapsychology*, W.G. Roll (ed.), Scarecrow Press, Metuchen, NJ, 1979.

39. Deborah L. Delanoy and Sunita Sah, 'Cognitive and physiological psi responses to remote positive and neutral emotional states,' in Dick Bierman (ed.), *Proceedings of Presented Papers, op. cit.*

40. C.T. Tart, 'Physiological correlates of psi cognition,' *International Journal of Parapsychology*, **5** (1963);
    Russell Targ and Harold Puthoff, 'Information transmission under conditions of sensory shielding,' *op. cit.*

41. Sybo A. Schouten, 'Applied parapsychology studies of psychics and healers,' *Journal of Scientific Exploration*, **7**,4 (1993).

42. Mario Varvoglis, 'Guérison psychique: Recherches experimentales et hypothèses théoriques,' *Revue Francaise de Psychotronic*, **2**,3 (1989).

43. —, 'Goal-directed- and observer-dependent PK: An evaluation of the conformance-behavior model and the observation theories,' *The Journal of the American Society for Psychical Research*, **80** (1986).

44. R. G. Stanford, 'Experimental psychokinesis: A review from diverse perspectives,' in B.B. Wolman (ed.), *Handbook of Parapsychology*, Van Nostrand Reinhold, New York 1977.

45. B. Millar, 'The observational theories: A primer. *European Journal of Parapsychology*, **2** (1978).

46. E.H. Walker, 'Foundations of paraphysical and parapsychological phenomena,' in L. Oteri (ed.), *Quantum Physics and Parapsychology*, Parapsychology Foundation, New York 1975.

PART THREE

SUPPLEMENTARY STUDIES

(I) Q V I DYNAMICS IN THE BRAIN

*by Attila Grandpierre**

The purpose of this note is to provide an assessment of the dynamics whereby human brains communicate with the scalar-mediated vacuum field. The findings presented here indicate that such communication is physically not only possible but actually plausible: the magnitudes it involves are consistent with known neural and physical processes, and they occur through known energy and information exchanges.

Assuming an energy transfer between the vacuum and quanta, and further assuming access of the quanta in the neural networks of the brain to fluctuations in the vacuum, we must account for an interchange of energy between the vacuum and quanta in the neurons of the brain. As Albert and Vaidman pointed out, the collapse of the wave function produces violations of the law of the conservation of energy, even if the amount of energy involved is small in macroscopic systems over all reasonable time spans.[1]

Here we first calculate the amount of energy released in the collapse of the wave function when information from an external source is accessed in the brain. The Heisenberg uncertainty principle tells us that quantum measurements cannot be arbitrarily precise in regard to both energy and time:

$$\Delta E \cdot \Delta t \geq h/4\pi \qquad (1)$$

---

* *Konkoly Observatory, Budapest*

The above law delimits the time-span of the total energy E of virtual particles to Δt. Virtual particles are created and annihilated with energy E for time Δt in accordance with

$$E \cdot \Delta t \approx h/4\pi \qquad (2)$$

Similarly the relationship between the position and momentum of virtual particles in the quantum vacuum is

$$\Delta p \cdot \Delta x \approx h/4\pi \qquad (3)$$

Equations (2) and (3) describe constraints on the physical properties of virtual particles.

We consider next that the quantum vacuum contains both vectorial (EM) waves and sub-quantum scalar waves. The latter may mediate finite energies below the threshold for the creation of real particles: $6 \cdot 10^{-27}$ ergs/sec. We may assume that the quantum wave function of a real particle collapses as it interacts with the energy carried by a scalar wave:

$$E(\text{scalar wave}) \approx \Delta E(\text{quanta}) \qquad (4)$$

Here for a real particle we note $p = mv$ and $E = p^2/2\,m$. For the sake of simplicity we adopt a coordinate system in which $p_0 = 0$, and $E_0 = 0$. We can then express position and energy together with momentum. Using m, a quantal carrier of mass as the basis of information, we obtain the following relationship between the overall size of the human brain and the time-scale of an information-accessing process in it:

$$\Delta x \approx h\Delta t / [8\pi m]^{1/2} \qquad (5)$$

For the human brain we take $\Delta x \approx 10$ cm, and, to satisfy the requirement for sufficient speed in mental processing to ensure the survival of the organism, give the time of an information accessing process as $\tau = 10^{-3}$ sec. [2] As the 'material carrier' of the process we take the electron, with mass $m_e \approx 9 \cdot 10^{-28}$ g. Given these values, we can estimate the lower limit of the size of macroscopic brain processes occurring via electrons as

$$L\,(\text{electronic brain}) \approx 10^{-2} \text{ cm} \qquad (6)$$

This limit is close to the size of cells, namely 10^{-3} to 10^{-2} cm. However, cells may use simultaneously a different material carrier in information processing, their environment being smaller and subject therefore to more rapid change. In the estimation of the relevant timescale we can note that the human organism seems able to process 10^9 to 10^{10} bit/sec information in nonconscious processes, whereas the conscious mind can cope with only 10^2 bit/sec. [3] If we make the reasonable assumption that the time-scale of the information processing in cells is proportional to the rate of nonconscious and conscious information processing, then the time-scale of cellular information processing is of the order of $\tau = 10^{-9}$ sec. [4] Inserting $\Delta x \approx 10^{-2}$

into equation (5) we get a value of 10^{-32} for the mass of this material carrier, which is proper for the mass-energy of a photon.

We can now estimate the wavelength of the information carrier waves. The energy of a wave quantum is E (wave) $= h\upsilon$, where υ is the frequency of the wave of a wavelength $\lambda = v/\upsilon$ and v is the propagation velocity of the wave quanta. If the latter have mass-energy $m = E/c^2$, with the help of equation (5) we get

$$m \approx \frac{h\tau}{8\pi(\Delta x)^2} \qquad (7)$$

Before interacting with the brain, however, the mass of virtual particles may be assumed to be negative (since the propagation of virtual scalars in the vacuum does not involve energy and is not constrained by the value of c [cf. Chapter Two]). Consequently for the virtual scalar waves v may be larger than the speed of light. The relationship between the wavelength of the information carrier wave and the spatiotemporal dimensions of the brain is thus

$$\lambda \approx \frac{8\pi(\Delta x)^2}{c\tau \approx \frac{2h}{(mc)}} \qquad (8)$$

For the human brain we take $\Delta x = 10$ cm and $\tau = 10^{-3}$ sec. This gives us $\lambda \leq 10^{-4}$ cm, which is close to the range of the visible spectrum. The magnitudes for cellular information processing are $\Delta x = 10^{-3}$ cm, and $\tau = 10^{-9}$ sec. The wavelength is $\lambda \approx 10^{-6}$ cm, a magnitude that is close to the ultraviolet range of the spectrum. (Interestingly, it is at this wavelength that, upon dying, the excess energy of the living cell is radiated off.)

The amount of energy expended in the collapse of the wave function can now be estimated, using the formula

$$\Delta E = \frac{p^2}{2m} \approx \frac{h^2}{32\pi^2 (\Delta x^2)m} \qquad (9)$$

In the case of the human brain, taking electrons as the material carriers of information, the relevant energy is $E \approx 2 \cdot 10^{-28}$ ergs. We can, however, look for more subtle information carriers. When electromagnetic light waves, rather than free electrons interact with vacuum scalar waves, the quantity of energy necessary to transfer information is less, proportionately to the difference between the mass-energy of wave quanta and the mass-energy of the electron

$$m(\text{wave}) = h\upsilon/c^2 = 2 \cdot 10^{-33} \text{ g} \qquad (10)$$
$$m(\text{wave})/m(\text{electron}) = 2 \cdot 10^{-6}$$

The smaller energy requirement in scalar/electromagnetic wave interaction means that, in the brain, vacuum fluctuations are more likely to affect electromagnetic waves than free electrons.

We now have a physical basis for describing the chain of events involved in the interaction between vacuum and brain. (In this description we take the transfer of information from vacuum to brain as the basis, noting, however, that the process from brain to vacuum proceeds precisely in the reverse direction.)

The information in question is contained in the vacuum scalar spectrum which is the brain's potential information source. In the *first transfer step*, some of this information becomes available to the neural circuits of the brain, as scalar wave patterns isomorphic with the Gabor transforms of particular neural networks interact with electromagnetic waves in the brain. In the *second transfer step*, the electromagnetic waves mediated by interaction with the scalars interact in turn with electrons and neurons in the brain. (This process is analogous to the transfer, through chlorophyll, of free energy from sunlight into plant biomass.) The chemical energies generated in the process build up the action-potentials of the given neural nets.

The scalar wave frequencies obtained here are remarkably close to the frequencies that occur in cell division. Macromolecules that exhibit an energy change of the indicated magnitude (10^{-28} erg) emit radiation at the wavelength of approximately 10 MHz. At the same time, Del Giudice *et al.* noted that the dynamics of biological processes may be driven by coherent electromagnetic processes that involve, as Jafary-Asl *et al.* observed, emissions in the radio frequency range of 7 MHz prior to the occurrence of cell-division.[5] This suggests that the interaction of scalar waves with electromagnetic information carriers in the brain may occur as a resonance phenomenon: the scalars transfer their energy and information to the brain's electromagnetic waves at coinciding frequencies. Resonance allows the transformation of the scalar waves of the vacuum into EM or material wave/particles without violating the energy conservation law at the macroscopic time scale. A resonant transformation process calls for the transfer of real energy only at its initiation: it is driven by the surplus of free energy available to the molecules of the living brain.

We have presented here a quantitative assessment of quantum/vacuum interaction in the brain. The process involves the transfer of information from scalar to EM waves and then to electrons and neurons, generating thereby chemical energies and modifying the cerebral processing of the accessed information.

References

1. D.Z. Albert and L. Vaidman, 'On a theory of the collapse of the wave function,' M. Kafatos (ed.), *Bell's Theorem, Quantum Theory and Conceptions of the Universe*, Kluwer, Dordrecht 1989.

2. H.B. Barlow, The languages of the brain, in *Encyclopedia of Ignorance*, 1977, 259-272.

3. Imre Törö, (ed.), *Az Élet Alapjai* (Foundations of Life), Gondolat, Budapest 1989, 106.

4. Ervin Bauer, *Elméleti Biologia* (Theoretical Biology), Akadémiai Kiadó, Budapest, 1967.

5. E. Del Giudice, S. Doglia, M. Milani, S.W. Smith and G. Vitiello, 'Magnetic flux quantization and Josephson behaviour in living systems,' *Physica Scripta*, **40** (1989) 786-791.

(II) PHILOSOPHICAL PERSPECTIVES OF QVI COSMOLOGY

*by Mauro Ceruti**

In the cosmological scenario of the Big Bang, it is possible to reconstruct, on the basis of our knowledge of physics, the history of the universe back to extremely remote ages in the past, even to times that are close to its origins. But it is the same physics that tells us that its laws do not hold for time zero, for that point of infinite density from which everything that exists in the cosmos had sprung forth. The origins of the universe must, it would seem, remain a singularity, an ultimate and unanalyzable limit of science's horizons.

The idea that at the origins of the universe there is an ultimate and unanalyzable singularity is consonant with cosmological concepts prevalent in the history of Western civilization. Indeed, from classical Greece up to the modern age, Western cosmology has pivoted around the idea of an ultimate and atemporal cause: a God who is before or beyond time, who creates time, and starts the Universe along its path of becoming. Scientists and philosophers have agreed not only in postulating an absolute beginning to the universe and to time, but also in proposing a moment in the development of human knowledge in which a stable congruence between knowledge and the universe will be reached. Today's project to build a 'theory of everything,' to explain 'all that is' through the unification of four universal physical forces is merely the last link in a long historical chain of cosmological speculation.

But different perspectives are now emerging. The beginning of *our* universe might not be the beginning of *the* universe—and of time and history. On the contrary, the Big

---

* *University of Palermo, Italy*

Bang (as a singularity and an a-dimensional point) might turn out to be a threshold with which to access other constraints and other regularities, other times and other universes—other histories.

In the early seventies Edward Tyson formulated the so-called 'hypothesis of the free lunch', a formulation, in terms of quantum physics, of the idea that the birth of the universe is due to a spontaneous fluctuation in a preexisting void; in the geometry of the spacetime that was to have existed before matter and radiation. In Tyson's cosmological scenario, the quantum vacuum is no longer pure nothingness. On the contrary, it is an entity that contains potentially innumerable virtual particles, a medium from which the particles are actualized and into which they are annihilated.

In the 1980s Ilya Prigogine, E. Gunzig, J. Geheniau, P. Nardone explored this conjecture in greater depth (cf. Chapter 3). They argued that during a timeless-time which was antecedent to the time of our universe, the quantum vacuum produced fluctuations of various kinds and intensities. One of these, of particularly great intensity, generated particles defined as the smallest black holes compatible with the known laws of physics and yet endowed with enormous mass compared to other particles. It is precisely this mass which triggered a series of nonlinear processes through which matter/radiation and the spatiotemporal structure of our universe were born, and through which they interact and shape each other.

This cosmological scenario introduces a distinction between the time of our universe, generated along with matter-particles in the initial creative instability, and the time of the quantum vacuum, the time in which the creative fluctuations themselves are produced. This is a radical amplification of what the historical and evolutionary sciences have put forward as the history of the universe. New forms of existence, autonomous ontological orders emerge in the world, and their emergence is accompanied by an irreducible specific temporality. Along with the universe's origins, today's physical cosmology seeks to explore these new dimensions relating to the ultimate destiny of the universe.

In regard to that destiny, two scenarios seem available. In the first scenario, the end of the universe is marked by the fire of the Big Crunch. In the second scenario, it is marked by the deep cold of an infinite void. But if the new hypothesis is valid, the end of *our* universe will not mean the absolute end of *all possible* universes. On the contrary, in the void that accompanies the dying universe, conditions similar to those which preceded its origins will be created—new fluctuations, new instabilities, new particles—a whole new universe. The irreversibility of events within our own universe is then inserted into a wider cycle.

The hypothesis of cyclical meta-universes increases the number of questions to which we wish to get answers. Would the new universes carry a trace, a memory of the past universes? Would their laws and physical constants differ from those of the foregoing universes? And if so, would they differ in a random manner, like a new turn on the cosmic roulette, or would their differences be in some way dependent on the history of the previous universes? And would they take in some way into account that previous universes were inhabited by living creatures? These questions demand an understanding of the history the cyclical meta-universe itself, above and beyond the history of our own universe. Ervin Laszlo dares to ask these questions—questions that project us beyond the confines of our local universe.

Yet the hypothesis of QVI (quantum/vacuum interaction) is much more than a development of multicyclic cosmology. The QVI scheme maintains that the quantum vacuum possesses more than its own reality; it possesses a reality that precedes matter and radiation in the cosmos. The vacuum plays a decisive role in the birth of our universe, and it continues to exercise a decisive role in its history and evolution, interacting with energy and with radiation, and producing a primary influence on the explorations and innovations of evolution. The vacuum field functions as a regulating and generative principle: it has a different, and in some sense more profound, actuality than matter and radiation.

This broadening of the cosmological picture is a decisive instrument for Laszlo's understanding of a whole series of anomalies, paradoxical experiences and surprising effects, such as quantum nonlocality, Pauli's principle of exclusion, the transmission of information among persons in distant places and cultures, and the collective unconscious as explored in Jungian psychology. He notes that there are different types of randomness, and those that appear to be absolutely random turn out to be the result of an interaction with a domain of reality that is much deeper than that which we have actually considered. There is no *absolute* randomness, only different degrees of contingency, relative to the scales and orders of magnitude in the connections that interlink different phenomena.

Laszlo's hypothesis is bold and controversial, and it is good science. More than that: it is a striking illustration of the positive feedback cycle between imagination and experimental control; a 'virtuous' rather than vicious cycle that is profoundly characteristic of the creative segments of the history of Western thought. In this history, radical and highly controversial hypotheses have been forged resting on metaphysical and mythical assumptions. But, at the same time, the hypotheses have been subjected to innumerable tests, confronted with a wide variety of facts and phenomena. The corroborations and falsifications which followed have enabled us discover much more than the hypotheses themselves had originally contained, shedding new light on their metaphysical and philosophical assumptions as well.

In expounding his hypothesis, Laszlo joins imagination with rigor. He has the daring of cosmological speculation, but also modesty of requiring continuous experimental testing for corroboration or falsification. As such, this study stands as a nucleating chapter in a collective history in which many voices and points of view come together, from different disciplines and divergent cultural contexts.

Indeed, many of the characteristics of the quantum vacuum, and of the hypothesis of the birth of the universe as a product of a giant fluctuation in a void which is beyond being and not-being, have antecedents that go far beyond the history of Western scientific thought. They recall distant cosmologies, which come to us from different times and places in the history of humanity. The Ether and the Bindu (initial-point) of the Hindu tradition, the Tao (the One) of the Chinese tradition, the Ein-Soph of the Hebrew Cabala, and the Apeiron (unlimited) of Anaximander recall the quantum vacuum in various ways, above all its aspect as a generator of potentialities. In a particularly poignant way, the Hindu vision revolves around a consistently multicyclic cosmology, in which the permanent and all-pervasive Absolute generates and reabsorbs universe after universe in the infinite immensity of cosmic eons. In all these cosmologies, the origins are not an absolute beginning from a single point in time, but rather a threshold which indicates the limit of current conceptual horizons, the place where the forms and distinctions that constitute the narrative of our universe fail, and where a common source appears from which these forms and distinctions themselves originate.

With this, as with so many scientific questions, we are beginning to recognize the acuteness and subtlety of knowledge conveyed by myths, rediscovering a new dimension of relevance reaching us from our own past as well as from other cultures. It appears that it is the entire cognitive experience of our species that has become pertinent to our present and to our future. In contrast with the traditional perspective of progress, in which those who are most up-to-date, with the most refined languages and the newest theories, surpass and erase all that has been produced before their time in the course of intellectual history, it is fruitful to have a history outlined here that connects and links us to problems and ideas that were cognized and recognized time and time again, in our own heritage as well as in distant cultures and traditions.

(III) NEW CONCEPTS OF SPACE AND TIME

The QVI scheme of a constantly interconnected and interactively self-evolving universe suggests new concepts of space and time. Since space and time are basic parameters of the physical universe, the indicated concepts merit further consideration. We begin with the concept of space.

The number of ways we can conceive of space is finite, and most of them have been explored in the history of science and philosophy. Without entering into undue detail — a comprehensive treatment of the subject would require a volume in itself — we can note the main alternatives.

To begin with, space can be considered an *a priori* concept added to experience by the rational mind; or it can be considered an *a posteriori* concept, a facet of experienced reality. Kant opted for the former, and Bohr among other physicists tended to agree with him. The same as Kant — and to some extent Heisenberg — Bohr held that the classical concepts of space and time are presuppositions of experience that are built into our everyday language; we can never escape them. He also agreed with Kant that the universe on the very large and the very small scale is not likely to fit concepts we have created to deal with our median-size world of experience. Though Bohr was sceptical of our ability to advance alternative ideas, contemporary mathematical physicists have no difficulty in coming up with abstract and non-experiential concepts such as Hilbert-spaces and superspaces, among many others.

The concept of space as an *a priori* condition of rational experience does not make it into a mapping of experience-independent reality; at best it offers a correspondence with such reality. However, taking the opposite tack, viewing space as an *a posteriori* concept, does not in itself ensure that it provides a one-to-one mapping of reality either. Indeed, in one of the *a posteriori* concepts—that of space as constituted by the relations between objects—we are free to deny the very existence of space: we can view it as the classical Void in which physically real entities (such as the atoms of Democritus) pursue their various careers.

We are, of course, equally free to assume that space as defined by relations between real bodies has an existence in the absence of those bodies. This is the view maintained among others by Newton, for whom three-dimensional space exists on its own. It exists, however, only as a passive backdrop: Newtonian space neither acts on material bodies, nor is it acted on by them.

But there is nothing to prevent us from viewing space as a real as well as an active element of the experienced world. Such a view amounts to a realistical interpretation of the topological concept pioneered by Poincaré. Following a suggestion of Minkowski (that space and time may be viewed as an integrated topological manifold) Einstein undertook to develop spacetime as an active matrix of the physical universe. In special relativity measurements of simultaneity and length are relative to the choice of inertial reference-frame within this integrated manifold. General relativity goes further—it identifies the metric of spacetime with universal gravitation. Here the spacetime manifold assumes a dynamical role: it interacts with material bodies. Its topology imposes an absolute value on the propagation of physical events, determines the shortest path between two points, accounting for gravitational attraction as well as for such relativistic effects as the slowing down of clocks and the contraction of measuring rods.

Except for the Kantian claims of some physicists, and the abstract mathematical concepts of quantum and grand unified theorists, relativistic spacetime remains the principal space-concept of contemporary science. It may not be its final pillar, however. Despite claims of universality, the spacetime of general relativity has encountered limitations. It does not hold either at very high energies or at very small distances: very energetic and very small interactions constitute 'singularities' in a relativistic universe. Einstein and his followers assumed that these are quirks in the theory and can be overcome, but more recently physicists, including Stephen Hawking and Roger Penrose, showed that the singularities are consequences of the theory itself. Moreover the relativistic limitation on the speed of light admits of exceptions. The most striking of them is the unexpected outcome of the thought experiment proposed by Einstein himself. The anomaly of the EPR experiment can be resolved, as John Bell suggested, by going back to the ideas of Lorentz and Poincaré, who believed in a preferred frame of reference. If we make that assumption, we can also assume that in the preferred frame things go faster than light.[1]

In regard to anomalously high-velocity signal propagation, the EPR result is not an isolated instance: it is joined by a growing number of phenomena where events are instantly or quasi-instantly correlated. These range from the Josephson effect, first noted in regard to superconductors and then discovered also in living systems, all the way to the time-and space-transcending phenomena investigated by Jungian and transpersonal psychologists.

Although all the facts are not in yet, it seems likely that physicists may have to face the challenge of a two-fold rectification of the relativistic spacetime concept. On the one hand, the manifold Einstein identified as space-time needs to be given physical reality, not just as a unitary geometry of space and time, but as a physical medium that fills the universe. On the other hand — and as a consequence of the above — space may again have to be distinguished from time. This may appear as a reversion to the classical concept, but it is not: the space to be envisaged is complex and active, not Euclidean and passive.

If the concept of time is disjoined from that of space, we shall have to consider it on its own. Time, of course, has been a problematic notion in physics ever since Newton failed to include Δt in the classical equations of motion. This has not been perceived as a flaw in most segments of the physics community: time, in the intuition of many physicists (and in this case Einstein is included) is a subjective phenomenon that can be 'transcended' in a mature theory. In this light the universe itself knows no time; it is only the observer that introduces it in virtue of his time-dependent observations.

The QVI scheme suggests that time is *both* a subjective *and* an objective phenomenon. It is certainly one of the key elements of our experience but, in addition, it is also a measure of events that occur independently of human observers. This affirmation is warranted in light of the history of evolutionary process thinking, first in the life sciences, and then in thermodynamics and cosmology. But it is not reflected in contemporary quantum theory, where Schrödinger's equations do not distinguish between the past and the future of a microscale quantum event any more than Newton's equations distinguish between the past and the future of a macroscale event. The indicated rectification may require, as Prigogine suggests, the introduction of large-scale Poincaré systems into the statistical description of quantum phenomena. In that case irreversibility—and hence time—will be introduced as a population property at the very foundations of the physical universe.

The concept of time indicated in light of the QVI scheme goes beyond this step, however. When time is introduced into the stochastic equations mapping quantum events, the time one obtains is not considered valid beyond the processes initiated with the Big Bang (or with subsequent 'bangs' in multicyclic cosmologies). For contemporary cosmologies, whether they embrace the standard scenario or a multicyclic one, the question "what was there before the instability that brought the universe into being" is not answerable—in terms of physics, it is meaningless. The fluctuations of the quantum foam that nucleate the observable universe are beyond time: they are random, and disconnected from the processes that result from their inflation.

Neither of these statements is necessarily true, however. We get nonrandom and interconnected quantum fluctuations if we allow that our universe arose in the 'in-formed' vacuum field of a concurrent open universe. The QVI scenario tells us that in its interaction the matter-energy component of the universe in-forms, and is in-formed by, its vacuum energy component. These two components of reality co-evolve, and the results of their co-evolution are registered in each of them. Thus the fluctuations of the quantum foam are no longer random: they are ordered by the quantum/vacuum interaction of an existing universe—regardless of whether that universe is ours, or came before ours.

The question, "what was there before the observable universe" becomes meaningful in terms of physical theory. Scalar-mediated ZPF fluctuations that had registered the Fourier transforms of micro- and macroscale systems in a precursor universe is 'what was there'. These fluctuations had originated in time, and they have persisted in time.

In this view time is coextensive with space. Both existed before our universe, and both will exist after it. Space and time are conjoined in actuality, but they are not integrated in theory; the integrated spacetime manifold is a geometrical *'Rechengrösse'* without physical reality. This concept could introduce a significant—and *ceteris paribus* highly desirable—measure of realism in the contemporary physical world picture.

Reference

1. John Bell, in P.C.W. Davies and J.R. Brown, eds., *The Ghost in the Atom*. Cambridge University Press, Cambridge 1986, pp. 48-49.

(IV) CREATIVITY, ARCHETYPES, AND THE COLLECTIVE UNCONSCIOUS

In a necessarily speculative but nonetheless reasoned vein, we shall consider here additional phenomena at the farther reaches of human experience. Can the phenomenon of creativity, and that of shared archetypes and the collective unconscious, be cogently viewed as effects of an interaction between the human brain/mind and the holofield of the quantum vacuum? We begin our explorations with the phenomenon of creative genius.

The question we pose first is whether acts of exceptional creativity occur in the isolation of a closed-system brain, or whether that brain—and the correlated mind and consciousness—may be effectively interacting with other brain-minds in the creative process.

Social and cultural influences on the minds of creative people are undisputed: no individual is a Robinson Crusoe, least of all sensitive persons such as creative artists, writers, composers, theoreticians, and others of their kind. However, the question raised here concerns a more immediate and spontaneous interaction than the standardly envisaged sociocultural influences. We raise the possibility that the minds of exceptionally creative people would be in spontaneous, direct, though not necessarily conscious, interaction with other minds within the creative process itself.

Subtle interactions beyond the scope of sensory perception have been suggested for millennia: they are an essential part of both Eastern and Western metaphysics and mysticism. As we have seen, in modern times many forms of ESP have been investigated in the laboratory, producing statistically significant results (cf. Chapter 5). "Twin pain" and image transference between physically related and emotively well attuned individuals, even when on distant locations, is relatively well established. The transactional and transpersonal schools of psychology acknowledge and research spontaneous subtle interactions between emotive and cognitive processes in individuals.

An interactive, transpersonal process could underlie exceptional acts of creativity as well. If so, certain forms of creative production may need to be traced to a confluence of interconnected cerebral processes, rather than to the brain and mind of one individual alone. The 'hypothesis of interactive creativity' could bring the astounding phenomena of genius closer to scientific understanding. In this note we first review the main strands of evidence relevant to the concept of interactive creativity, and then suggest that the universal interconnections described in the QVI scheme can provide a potentially fruitful interpretation of the facts.

The tenet of interactive creativity suggests that some acts of creativity and some insights are due not to a spontaneous and largely unexplained stroke of genius, but to the elaboration of an idea or a pattern in two or more minds in interaction. This is equivalent to a dialogue in the Platonic sense of the term—to a process in which the results transcend the individual abilities of the partners in the dialogue. It also recalls Plato's view that in the course of an insightful dialogue the Soul 'recollects' the key ideas...we need only to substitute 'collection' for 'recollection'. According to the hypothesis of interactive creativity, in the course of the creative process persons *collect* some elements of the product of their creativity from other persons, rather than *recollect* them from a higher dimension of Forms and Ideas.

Independent evidence suggests that genuine acts of creativity are often based on an *Einfall*, a sudden and spontaneous intuition leading to a conceptual or aesthetic breakthrough. The appearance of such intuitions is not well understood. Individuals known for repeated and remarkable *Einfälle*—a Mozart, a Michelangelo, or a Shakespeare, to name but a few—are regarded as geniuses, born with rare and mysterious gifts. This view is reinforced by the fact that sometimes otherwise entirely unremarkable individuals display astonishing, seemingly inborn, capacities in specific fields, especially in music and in mathematics. To call such individuals 'gifted' and their achievements 'works of genius' is not to explain their abilities, but only to label them. An explanation involves answering questions regarding the origins of their unusual accomplishments. Are they possessors of a specially fortunate combination of genes? Or did they receive their gifts from a higher source?

Questions such as these do suggest answers, but they are not scientifically acceptable. Yet scientifically acceptable explanations may be possible. We should note first of all that some of the most remarkable *Einfälle* occur in altered states of consciousness. Few artists compose music and poetry or paint and sculpt in an ordinary common-sense frame of mind. There is almost always some element of transport to another plane of consciousness, a deep concentration that approaches a state of trance. In some (relatively rare) cases these 'inspired states' are artificially induced—by drugs, music, self-hypnosis or other means. Mostly, however, they come spontaneously to the 'gifted' individual. Coleridge composed his celebrated epic poem *Kubla Khan* while lying in what he described as a profound sleep (which was in fact induced by laudanum, an opium-based substance he took as medicine); Milton created his *Paradise Lost* as an 'unpremeditated song' dictated, he said, by the Muse. Mozart claimed that his compositions came to him during nights when he could not sleep. They came complete, from where he could not fathom. He did not hear the parts one after another, but the whole piece at once. "What a delight this is," he wrote in an oft cited letter, "I cannot tell. All this inventing, this producing, takes place in a pleasing lively dream."

In the sciences, too, altered states occur with remarkable frequency in the course of innovation and discovery: many scientific breakthroughs owe their existence to an altered state of consciousness. This is true of mathematical discoveries as well. Evariste Galois, for example, committed to paper his fundamental contributions to higher algebra at the age of twenty in three feverish days before meeting an adversary in a duel he expected to be fatal (and which was that indeed). Karl Friedrich Gauss sought to discover the proof for the way every number can be represented as the product of primes and, though he made many tries, did not succeed for years. After many failures he could write in his diary that he had succeeded, "but not on account of my painful efforts. Like a sudden flash of lightning, the riddle happened to be solved." Henri Poincaré said with good reason that the elements of a mathematical discovery are "harmoniously disposed so that the mind without effort can embrace their totality—divining hidden harmonies and relations."

Perhaps, exceptional achievements in both science and art can be given rational explanation through the hypothesis of interactive creativity. Consider the relevant parameters of the creative act. A person with a high level of motivation and great powers of concentration focuses on a given task or problem. Another person, likewise highly motivated and concentrated, focuses on the same or a closely similar task. In these conditions the similarity of the states of brain and mind in these individuals allows some level of access to each other's cerebral processes. This gives rise to a subtle dialogue that often has remarkable consequences. In this dialogue the individual may receive a sudden flash of insight, sufficient to start off the process of creative elaboration, or may find himself 'guided' by a sustained though subconscious process during which now this avenue and now that is explored, assisted by flashes of intuition. These are relatively frequent experiences not only in artistic creativity, but also in the context of scientific explorations (the present writer has been fortunate to experience it on several noteworthy occasions).*

The above thesis is supported by the fact that, as discussed in Chapter 5, when different persons bent on a related creative task enter a state of deep concentration, their brain states become highly synchronized. This occurs whether or not they are physically on the same location. It could occur also whether or not they consciously know of each other. In fact, the absence of conscious awareness is likely to be an advantage in a creative process, as at least in Western individuals normal waking consciousness is usually dominated by the linear logic of the left cerebral hemisphere, and such logic tends to suppress unusual contents of consciousness. Altered states of intense concentration (or meditation) are relatively free of such censorship. They allow subtle inputs—*Einfälle*—to fertilize and inspire creative endeavors.

When elaborated in theory and subjected to empirical scrutiny, the hypothesis of interactive creativity may overcome the spurious alternatives of ascribing faculties of genius either to gifts, or to genes. It could show that a better understanding of the phenomenon involves reference to spontaneous interconnections in the realm of mind and consciousness: constant, if largely unconscious, linkages between the cerebral processes of similarly orientated human beings. A recognition of these linkages could bring the as yet speculative domain of the philosophy of creativity within the compass of empirical science.

---

* Interactive creativity has a functional analogy in the functions of networked computers. When their work stations are linked within a computer network, users can access and elaborate data created by other users on other work stations. Items created by several individuals can enter the same document, and each individual can add further elements to them. Ordinarily, users will know which item they have entered themselves, and which came from a networked source. In the creative process, however, such rationality is excluded: individuals are not aware of being linked up with other individuals, nor are they in a state of mind conducive to tracking phenomena of a seemingly extrasensory variety. All that they are aware of is the 'falling in' (*Einfall*)of an idea or notion, which then functions as a trigger for their own creativity.

There are other forms of spontaneous communication in human experience—forms that link the human brain/mind with entire cultures and societies, as well as with the natural environment. It is to these phenomena that we turn next.

Social and cultural anthropologists observed that in well functioning social systems the individual members are thoroughly integrated in the social fabric. How such integration can come about is not clear. Do people become integrated in their social time and place merely by perceiving how others behave and learning to behave the same way, or is there something in their genetic make-up that predisposes them to common forms of thinking and behavior? Is, in other words, the phenomenon of human socialization based exclusively on 'nurture' or is it also prompted by 'nature'? The latter alternative is favored by geneticists, who believe that genes may also code emotional and moral behavior—a thesis keenly contested by the majority of behavioral and social scientists. The former alternative is usually espoused by social scientists, yet the 'nurture' thesis is not unproblematic either: whether learning through language and the emulation of perceived behaviors can fully account for the social sense and morality displayed by individuals is questionable as well. It is possible, then, that the brain/mind of individuals is affected, in subtle yet significant ways, by the milieu in which they are integrated, much as the genome is so affected (cf. Chapter 4). Social empathies and cultural identifications may require a deeper root in the psyche of the individual than learning through language and emulation can provide.

It may be that the coherence and adaptation of a sociocultural system rests on a field-mediated coordination of individual humans, much as the coherence of the living organism rests on the Josephson effect, a likewise field-mediated coordination of cellular assemblies. Indeed, brain-holofield interaction may have been the principal cohesive factor in the epoch which intervened between the formation of tribal structures and the development of linguistic communication. (If linguistic abilities appeared, as some theorists maintain, not more than 50,000 years ago, then the prior epoch of sign- rather than symbol-based socialization would embrace several hundred thousand years.) The widely observed natural empathies of so-called 'primitive' peoples reinforce this supposition. To cite merely the oft-quoted nineteenth century declaration of Native American Chief Seattle: "This we know. All things are connected like the blood which unites one family. All things are connected. Whatever befalls the Earth befalls the sons of the Earth."

In Western and westernized societies natural empathies are mainly repressed; they surface principally in the mind of exceptionally sensitive individuals. Poets such as John Donne and William Blake have sung of our oneness with nature, and a handful of scientists, such as William James, Abraham Maslow, Gregory Bateson, and Arne Naess, have sought a detailed understanding of it. But the consciousness of the typical individual is rarely affected by such empathies—possibly a major reason why modern societies struggle with incomprehension in the face of increasing environmental deterioration.

Feelings of oneness with the environment may be more than the remnants of a primitive mentality, or the subjective idiosyncrasy of a few poets and scientists: their roots may lie in the interaction of the brain with the vacuum holofield. According to the QVI scheme the pattern the brain accesses from that field is multidimensional: it includes the traces of both the subordinate and the superordinate levels of the systems in which the individual is embedded. In one direction the pattern would include the Fourier transform of the configuration of cells and molecules that constitute the human body; while in the other it would include the dynamic $3n$-dimensional configuration-space of the social and ecological milieu of the individual.

Not only individuals, entire cultures can enter into spontaneous communication with each other. Here the evidence is historical and thus not reproducible, but it is nevertheless impressive. For example, the great breakthroughs of classical Hebrew, Greek, Chinese, and Indian culture occurred almost at the same time, though they occurred in widely scattered regions among people who were not likely to have been in actual communication with each other. The major Hebrew prophets flourished in Palestine between 750 and 500 BC; in India the early Upanishads were composed between 660 and 550 BC and Siddharta the Buddha lived from 563 to 487 BC; Confucius taught in China around 551-479 BC; and Socrates lived in Hellenic Greece from 469 to 399 BC. At the time when the Hellenic philosophers created the basis of Western civilization in Platonic and Aristotelian philosophy, the Chinese philosophers founded the ideational basis of Oriental civilization in the Confucian, Taoist and Legalist doctrines. While in the Hellas of the post-Peloponnesian Wars period Plato founded his Academy and Aristotle his Lyceum, and scores of itinerant sophists preached to and advised kings, tyrants and citizens, in China the similarly restless and inventive 'Shih' founded schools, lectured to crowds, established doctrines, and manoeuvred among the scheming princes of the late Warring States Period. Space-leaping phenomena of this kind is either a sociocultural variant of action-at-a-distance, or it occurs by means of a medium of transmission. The QVI hypothesis—as any potentially scientific explanation— suggests the latter.

Phenomena of 'cultural synchronicity' require a naturalistic interpretation of the notion of archetypes. In this interpretation archetypes, and the collective unconscious that frames them, are not just 'in the mind': they are 'in nature'. Insightful if speculative thinkers such as John White tell us that archetypes and the collective unconscious constitute a subtle envelope surrounding the planet; a shell to which people have access during dreams, meditation, and other altered states of consciousness. Here resides the Akashic record that Edgar Cayce and other psychics claim to 'read' when they obtain psychic information.[1]

In Carl Jung's more orthodox view, archetypes arise from a vast, limitless unconscious process shared by all humanity, emerging from the accumulated experience of thousands of years of shared history. Initially Jung speculated that archetypes are due to the gradual modification of genetic structure in individuals, allowing personal experience to incorporate ever more elements of the collective unconscious. Later, however, he gave up trying to give a physiological explanation—the mechanistic brain science of his time was not ready to acknowledge spontaneous communication between people and culture groups on distant locations. In the commentary to *The Secret of the Golden Flower* Jung reaffirmed, however, that "the collective unconscious is simply the psychic expression of the identity of brain structure irrespective of all racial differences."[2]

Shortly before his death in 1961, Jung ventured even further. "We might have to give up thinking in terms of space and time when we deal with the reality of archetypes," he wrote in a letter published in the same year in German. "It could be that the psyche is an unextended intensity, not a body moving in time. One could assume that the psyche arises gradually from the smallest extension to an infinite intensity, and thus robs bodies of their reality when the psychic intensity transcends the speed of light. Our brain might be the place of transformation, where the relatively infinite tensions or intensities of the psyche are tuned down to perceptible frequencies and extensions. But in itself," Jung added, "the psyche would have no dimension in space and time at all."[3] Jung's disciple Marie-Louise von Franz concurred. The psyche is eternal and everywhere, she asserted. When something that touches it happens at one point, it has happened at all points simultaneously.[4] A similar view was voiced by Stanislav Grof. "Modern consciousness research reveals" he wrote, " that our psyches have no real and absolute boundaries; on the contrary, we are part of an infinite field of consciousness that encompasses all there is—beyond space-time and into realities we have yet to explore."[5]

The unified QVI scheme refers the collective, space- and time-dimensionless psyche addressed in such speculations to a general and fundamental feature of the universe: the persistent interaction between the holofield of the quantum vacuum, and the micro- and macroscale systems that evolve in space and time.

References

1. John White, *The meeting of Science and Spirit*, Paragon House, New York 1990.

2. Carl G. Jung, Commentary on *The Secret of the Golden Flower*, in R. Wilhelm, *The Secret of the Golden Flower*, Harcourt, Brace & World, New York 1962.

3. quoted by Marie-Louise von Franz, *Psyche and Matter*, Shambhala, Boston and London, 1992, 161.

4. *Ibid.*

5. Stanislav Grof with Hal Zina Bennett, *The Holotropic Mind.* Harper San Francisco, 1993.

(V) PROBLEMS AND PROSPECTS OF TRANSDISCIPLINARY UNIFIED THEORY

As the here presented concepts should have demonstrated by now, theories that attempt a transdisciplinary unification of our understanding of physical, biological and psychological phenomena create a fundamental change in the way we look at ourselves and the world. The most basic presuppositions of disciplinary theories undergo subtle yet significant alteration. This process has been well described in the literature on paradigms: a new paradigm alters the way we look at data, changing our most basic assumptions about the nature of the phenomena under study. In the case of transdisciplinary unification of the evolutionary variety, a specific worldview-factor intervenes. In adopting an evolutionary framework we no longer look at phenomena as things or objects that are to be described as they *are*. Instead, we describe phenomena in terms of how they *come to be*. Paraphrasing Einstein, in the unified evolutionary context "we are seeking for the simplest possible scheme that can explain how the facts have been *generated.*"

Seeking that scheme may run into practical difficulties. Contemporary science, as observers have often noted, still has a disciplinary bias. Unification is sought mainly within the boundaries of individual disciplines, and researchers in competing academic departments view with distrusts attempts by investigators outside their specialty domain to encroach on their territory. However, distrust in regard to unified theories of the evolutionary variety is entirely misplaced. Even the briefest look at the history of the concept of evolution testifies that the territoriality that is the habitual correlate of scientific investigation does not apply here. Evolution does not 'belong' to any given field of study; it is the common domain of all the empirical sciences.

The term itself comes from the Latin *evolvere*, meaning to unfold. It was first applied, erroneously as it turned out, to the development, or 'unfolding', of the full-grown organism from the minute homunculus that was presumed to exist, fully formed, in the male sperm or the female egg. Then, in the nineteenth century the concept of evolution came to be identified with the theory of Darwin and the field of macrobiology.

Many scientists, first and foremost biologists, still lean toward the view that evolution is a biological concept, without direct or significant application in other fields. Yet, in the light of current findings, such preconceptions are groundless. We need merely to recall that, out of the probabilistic, quantized ground investigated in field physics emerged in succession the more measurable and knowable phenomena of the macroscopic world. The synthesis of matter began in the first 10^{-33} second that marked the end of Planck-time; and the synthesis of progressively heavier elements continued ever since, both in stars and in interstellar space. Biological evolution took off on this planet sometime between 3.6 and 4.6 billion years ago; hominid species appeared in the last few million years with sapiens emerging as a species capable of language, tool-use and abstract thought in the span of the last 100 thousand years. And some 20 thousand years ago the tribal communities of Homo, coded by the 'communicational realities' brought about by the enlarged cranial capacity of the species, began to evolve into today's complex sociocultural and technological societies. Despite its probabilistic and problematic basis in quantum fields, the world we observe today worked itself up on Earth, and possibly elsewhere in the universe, from common and presumably unified beginnings to its present diversified—but not unordered—condition.

Yet it would be a mistake to look at the concept of evolution as a grab bag for concepts of change in nature. It is only on a superficial level that it may seem that there is no process of change that would be meaningfully excluded from the idea of evolution; on closer inspection many exclusions become evident. First of all, although evolution is as theory of change, it is not the theory of all varieties of change. Purely random and entirely time-reversible patterns of change are excluded: evolution concerns exclusively change that is, at least statistically, irreversible. But not even all forms of irreversible change fall within its compass. To qualify, irreversible change must entail processes that lead to the emergence, or at least the persistence, of ordered structure in space and time. Such processes constitute an orderly sequence that must prove traceable from the origins of the physical universe through multiple hierarchical levels and processes to whatever state or process we wish to consider. Evolution is the study of progressive and ongoing, but not necessarily continuous and linear, and by no means fully predictable (though logically retrodictable), change, leading with a statistical irreversibility from the origins of the cosmos to its present and future states.

With the advent of nonequilibrium thermodynamics, and Big-Bang as well as multicyclic cosmologies, the statistical irreversibility that is the hallmark of evolutionary change has entered the theories of contemporary physics—it is no longer the exclusive preserve of the life sciences. Irreversibility is discovered also in the social sciences, above all in history and in psychology. Aside from theories of cyclical and eternal recurrence, and positivist disclaimers of the meaningfulness of large-scale patterns, many social scientists appear inclined to make use of the concept of sociocultural (or socioeconomic or sociotechnological) evolution. Similar notions have surfaced in psychology and personality theory, where investigators such as Piaget and Kohlberg have thrown light on the sequential and apparently irreversible unfolding of perceptual, intellectual and moral faculties in the life of the individual.

That evolution unfolds sequentially does not mean that it must be continuous. Discontinuities exist between physical-chemical, biological, and human and social phenomena—they have been the main factors responsible for the persistent separation of science's disciplinary fields of investigation. Such compartmentalization has led to a one-sided emphasis of differences among the investigated phenomena at the expense of parallelisms, isomorphies, and functional analogies. As a result today we have numerous, highly specialized and independently conducted studies of evolution—studies of the *evolution of* particular entities, such as stars, butterflies, cultures, or personalities, but few if any truly unified concepts of evolution as a fundamental process.

Yet the disciplinary research of the *evolution of* various phenomena need not prove to be a lasting obstacle to the creation of a truly unified evolution theory. In the last decade of this century the kind of disciplinary separation that characterized the opening decades promises to be overcome. It is becoming evident that processes in the physical and in the living world are not 'complementary' in the sense that Bohr intended in the 1920s when he argued that it is just as impossible to bring both these processes into sharp theoretical focus as it is to specify both the wave-like and the particle-like aspects of the quantum. Since the physical and chemical properties of the substances that make up living systems are physical in origin, it would be paradoxical to maintain that the laws that govern the parts are basically and incurably incompatible with the laws that regulate the whole. Rather, the question is whether the *formulation* of these laws, as they apply to the physico-chemical parts, are adequate to grasp the processes that are manifested by the complex wholes constituted of these parts. This raises questions concerning current interpretations in quantum mechanics, thermodynamics, genetics and biological evolution theory. If in these theories mutual incompatibilities appear, they are more reasonably ascribed to flaws in the interpretations than to inconsistencies in nature. If one cannot assume that nature is intrinsically divided into separate compartments, one must not insist that its investigation should be conducted within compartmentalized disciplinary frameworks.

The program of formulating a transdisciplinary unified theory is intrinsically coherent. In some of its aspects, it is part and parcel of a stream of scientific innovation that began in antiquity. In one of these aspects, it sets forth science's increasingly penetrating probings of the fine-structure of the universe. The indivisible atom of Democritus was rediscovered by Dalton and Lavoisier as the basic constituent of gaseous matter, but their atom proved divisible. Hence the floor of inquiry was lowered to the level of the particles that surround the nucleus of Rutherford's atom. A deeper basement has then been reached with the discovery of the quantized nature of light: inquiry then moved to the level of Planck's constant. The advent of quark theory bored physics' leading edge even further into the microworld. Not surprisingly, attention came to be focused on the quantum vacuum, the domain in which these progressively more minute and abstract entities are embedded. Space itself transformed from the passive Euclidean concept espoused in classical mechanics to the turbulent energy-filled vacuum of the new physics—and it may further transform into an interactive fifth field in nature.

The lowered floor of scientific inquiry has extended the effective range of scientific knowledge. In the seventeenth century Galilean physics described mechanistic processes on the surface of the Earth, and subsequently Newtonian mechanics expanded the range of these descriptions to all bodies moving within inertial frames. At the beginning of this century Einstein extended the validity of physical laws to accelerated frames up to the speed of light, and two decades later Bohr extended the laws of physics to the subatomic world. It now appears that relativity physics holds good only to about 10^{-8} m and that quantum physics, though it claims validity all the way to the Planck-length of 10^{-35} m, encounters anomalies (such as those associated with the energy density of the vacuum) at the level of 10^{-20} m. But if transdisciplinary unified theory shifts the roots of inquiry from the level of quanta to that of the field that interconnects them, science will extend the range of theory construction and experimentation still further.

As these canonical trends continue in the twenty-first century, we can foresee the phase in the development of natural science when its presently still discipline-bound investigations become consolidated through the mathematical formulation of the transdisciplinary dynamics that drives evolutionary processes in the diverse realms of observation. Given that evolution knows no disciplinary boundaries, the emerging transdisciplinary unified theory will describe the various phases and facets of the evolutionary process with invariant general laws. The laws will enable investigators to describe the behavior and evolution of quanta, and of the atoms, molecules, cells, organisms, and systems of organisms that evolve from quanta, in reference to a self-consistent, mathematically formulated and transdiscipinarily unified scheme.

The equations of the mature theory will have the form $D\ F\ (i,j)$, where a universal integral-differential operator D defines the general phase-space density $F(i,j)$, with the i,js representing the generalized positions and momenta of micro- and macroscale systems in the quantum vacuum. The latter will constitute a universal reference frame for the computation of the position and motion of all directly and indirectly observable phenomena.

AFTERWORD*

by Karl Pribram

The Interconnected Universe is a superb example of postmodern deconstruction at its very best. It demonstrates the anomalies and lacunae in the current narrative we call science and develops a new narrative that aims to carry our comprehension beyond these limitations. In this regard I would caution the reader to adhere to a maxim once issued by Warren McCulloch: 'Do not bite my finger; look where I am pointing.'

The term 'narrate' is closely linked to the Latin '*gnarus*' which in turn is kin to '*gnoscere*', to know. Thus narration is a form of knowing, just as is 'science,' *scire*, kin to '*scindere*', to cut. However, the accepted language of science is mathematics, a sharply honed tool for cutting, analysing, observations into packets that allow observations to be shared (replicated). Paradoxically, mathematics also allows predictions to be made, predictions which lead to new observations.

Twentieth-century science has been eminently successful in its pursuit of *scire*. Cognitively, however, mathematical formulations are, by themselves, incomplete. The narrative aspects of science, the concepts and meanings to which the computations point, have been neglected, often deliberately as in the ever-popular Copenhagen interpretation of quantum physics. This neglect has produced considerable malaise in some of us; and

---

* Revised version of a Foreword originally published in Ervin Laszlo, *The Creative Cosmos*, Floris Books, Edinburgh, 1993.

more important, it has led to a cover-up of the anomalies and lacunae addressed in the present study.

The Interconnected Universe ably summarizes what is missing in today's account of science-as-narrative. Of course, Laszlo is not alone in his lament. Einstein, Dirac, Bohm and Bell have all attempted to understand their formulations in physics; Koestler, in biology and psychology. But the received wisdom in the classroom has, for the most part, emphasized the elegance of what has been achieved often with the advice that any attempt at further understanding would simply confuse. Laszlo is to be commended in that he provides us with a plausible alternative. All of the scientists noted above have groped in the direction now taken by Laszlo. He points out that, as the twentieth century comes to a close, scientists are again becoming more comfortable with the concept of 'field' which has been eclipsed for most of the century by an almost exclusive emphasis on the particulate.

Fields are invoked to account for (inter-) actions at a distance. Newton conceptualized such actions and interactions in terms of force. Today we have become so accustomed to this innovation that we think of the force of gravity as a thing. Actually, of course, all we have are the observations of actions—interconnections—at a distance. As Laszlo indicates, this means that we are inferring gravity from our observations: gravity is not an observable; as in the case of field concepts in general, it is inferred.

The reason why we have become so accustomed to the inference that gravitational fields exist, is that there properties have been clearly stated and shared observations have made them common cognizance. Of course, these commonalities hold only for two interacting bodies; the three body problem is only now beginning to yield to nonlinear computations.

Gravitational, electromagnetic, the strong and weak nuclear forces have all become relatively familiar, because their inferred properties do not invoke any radical departure from the measurements that have served scientists so well. These four fields are inferred from interactions among entities. The interactions take place in space and over time. Probabilities of occurrence must be invoked for the nuclear forces; the inverse square law is replaced by the Pauli exclusion principle, etc., but on the whole the inferences are manageable, albeit barely.

The postulated interaction of Laszlo's QVI scheme is different. It is not inferred from an interaction among spatially and temporally separated entities. As Bohm has described

AFTERWORD

145

it, space and time become implicate, enfolded. Mathematically, the interactive field is spectrally, holographically organized. The organization is composed of interference patterns, that is, of the amplitudes (amounts) of energy present at intersections among waveforms. The equations that describe the transformations from spacetime to spectrum are called spread functions because the changes in form spread entities into a distributed manifold of such amplitudes. The relations among amplitudes can be conceived to compose a holoscape which can be represented by a contour map, similar to the familiar weather map of temperature gradients.

Laszlo's interactive quantum field is thus not a simple inference from observations. Rather, it is a transformation of fields, which are inferred from observations. It is this second order aspect of the interconnecting field which makes it difficult to grasp. In fact, until the engineering instantiation of the holographic mathematical formulation, only mathematicians were able to imagine this type of organization. Leibniz was the first in describing what he called monads; Gabor was responsible for holography in the immediate past.

Scientists do not feel as comfortable with transformation as do mathematicians. Transformations imply action paths. Until recently, scientists have conceived of action paths mostly in terms of conservation laws, optimizations in favor of least action principles. Prigogine broke ground in pointing out the difference between stabilities that occur far from equilibrium and those that reflect least action, i.e., equilibrium. The paths described by the transformations between spacetime and spectrum and back again may lead to stabilities far from equilibrium as well as to least action. When least action is involved, the path (integral) has the form of quantum mechanics. Gabor showed that this form is not limited to the realm of quantum physics but is universal to communication. He therefore defined a quantum of information, a channel which can carry a unit of communication with the least amount of uncertainty.

The path to least uncertainty leads, of course, towards communication with maximum amount of information. By using the measure of the amount of information as a measure of diversity within some domain (that is, a measure of complexity), we come directly to Prigogine's description of the path to order from chaos.

There are thus at least two classes of paths that describe the transformations that occur between the spectral and the spacetime domains. Both begin by describing a phase space that combines the spectral and the spacetime domains. One class emphasizes the spectral aspect as it describes the spatial composition of matter; the other emphasizes the spacetime aspect as it describes the time-evolution of complexity.

There is one aspect of the QVI scheme which needs to be especially carefully considered. As a physiologist, I know that the same or similar descriptions (even mathematical formulations) hold at various levels of organization. Some aggregates of molecules behave much as do some aggregates of cells—and even some aggregates of individuals. Thus laws of economics may be applicable to the operations of neural systems; and the feedback loops involved in homeostasis and homeorhesis are as applicable to control procedures in engineering as they are to the regulations of an organism's internal environment. Telephone communication systems used as information transmission networks and the programming of computers make elegant metaphors for understanding how the brain operates to generate and control behavior and experience. But the nitty-gritty of the use of these metaphors is the construction of precise models that detail the similarities and differences in transformations, the transfer functions that describe *how* different levels influence one another.

I have formulated the hypothesis, based on evidence quoted by Laszlo, that hyperstimulation of the frontolimbic forebrain allows primates including humans to operate in a holistic, holographic-like mode. I have trouble imagining *how* the interactive vacuum field would influence the organism in any detailed and specific extrasensory fashion. In the same vein, I have difficulty attributing to that field the permanent human-type memory storage capability envisioned by Laszlo. As I conceive it, the field enters the transformation process—that is, it describes paths of transformation at *several* levels or scales—thus obviating the need to overarch from cosmos to quark. But these may be my own limitations. When I was a child, I had trouble believing that heavier-than-air craft would be capable of mass transportation, dirigibles made much more sense to me. I am still awed by portable radios and their current walkman size. Television leaves me gaping in color shock and the proximity of faces. Thus, my read of *The Interconnected Universe* reminded me of treasured accounts of adventures, such as *The Microbe Hunters*, and Admiral Byrd's exploration of Antarctica.

Laszlo has, indeed, filled the need for a twenty-first century renewal of the narrative of science which has been so neglected during the twentieth.

BIBLIOGRAPHY

Adler, T., 'Meta-analysis offers precision estimates.' *APA Monitor*, September 1990.

Albert D.Z. and L. Vaidman, 'On a theory of the collapse of the wave function,' *Annalen der Physik*, **35** (1911), 898-908.

Barlow, H.B., 'The languages of the brain,' *Encyclopedia of Ignorance* 1977, 259-272.

Barrow John D., and Frank J. Tipler, *The Anthropic Cosmological Principle*, Oxford University Press, London and New York 1986.

Bauer, Ervin, *Elméleti Biológia* (Theoretical Biology), Akadémiai Kiadó, Budapest 1967.

Bearden, Thomas E., *Toward a New Electromagnetics*, Tesla Book Co. Chula Vista, CA, 1983.

Beauregard, O. Costa de, *Le Temps Déployé*, Editions du Rocher, Monte Carlo 1988.

Benor, Daniel J., *Healing Research: Holistic Energy Medicine and Spiritual Healing*, Helix Verlag, Munich,1993.

Bohm, David, *Wholeness and the Implicate Order*, Routledge & Kegan Paul, London 1980.

Bohm David, and B.J. Hiley, 'Non-relativistic particle systems.' *Physics Reports* **828** (1986).

Bohm David, and B.J. Hiley, *The Undivided Universe*, Routledge, London 1993.

Bohr, Niels, in *Albert Einstein, Philosopher-Scientist*, P. Schilpp (ed.), Cambridge University Press, London 1970, 205-206.Braud W., and M. Schlitz, 'Psychokinetic influence on electrodermal activity. *Journal of Parapsychology*, **47** (1983).

Byrd, R.C., 'Positive therapeutic effects of intercessory prayer in a coronary care population,' *Southern Medical Journal*, **81.7** (1988)

Costagnino, Mario, Edgard Gunzig, Pascal Nardone, Ilya Prigogine and Shuichi Tasaki 'Quantum cosmology and large Poincaré systems (mimeo), 1994.

Davies, P.C.W. and J.R. Brown (eds.), *The Ghost in the Atom.*, Cambridge University Press, Cambridge 1986, pp. 48-49.

Dean, E.D., and C.B. Nash, 'Coincident plethysmograph results under controlled conditions,' *Journal of the Society of Psychical Research*, **44** (1967).

Del Giudice, E., G. Preparata and G. Vitiello, 'Water as a free electric dipole laser,' *Physical Review Letters*, **61,9** (1988).

Del Giudice, E., G., S. Doglia, M. Milani, and G. Vitiello, in F. Guttmann and H. Keyzer (eds.), *Modern Bioelectrochemistry*. Plenum, New York 1986;

Del Giudice, E., *Nuclear Physics*, B275 (FS17), **185** (1986).

Del Giudice, E., S. Doglia, M. Milani, C.W. Smith and G. Vitiello, 'Magnetic flux quantization and Josephson behaviour in living systems.' *Physica Scripta*, **40** (1989).

Delanoy, Deborah L., and Sunita Sah, 'Cognitive and physiological psi responses to remote positive and neutral emotional states,' in Dick Bierman (ed.), *Proceedings of Presented Papers,* American Parapsychological Association, 37th Annual Convention, University of Amsterdam, 1994.

Detlefsen, Thorwald, *Schicksal als Chance*, Bertelsmann Verlag, Munich 1979;

Duke D.W. and W.S. Pritchard (eds.), *Measuring Chaos in the Human Brain,* World Scientific, London and Singapore 1991.

Eccles John, and Daniel N. Robinson, *The Wonder of Being Human,* Shambhala Publications, London 1985.

Edelman, Gerald M., 'Morphology and Mind: is it possible to construct a perception machine?' *Frontier Perspectives,* **3**,2 (1993).

Edelman, Gerald M., *Bright Air, Brilliant Fire On the Matter of Mind,* New York, Basic Books, 1992.

Einstein, Albert, *Über die spezielle und die allgemeine Relativitätstheorie*, Akademie Verlag, Berlin 21. Auflage 1916.

Einstein, Albert,*The World As I See It*, Covici-Friede, New York 1934.

Einstein, Albert, 'Einfluss der Schwerkraft auf die Ausbreitung des Lichtes', *Annalen der Physik*, **35** (1911).

Eldredge, Niles *Time Frames: The Rethinking of Darwinian Evolution and the Theory of Punctuated Equilibria*, Simon & Schuster, New York 1985.

Eldredge, Niles, and Stephen J. Gould, 'Punctuated equilibria: an alternative to phylogenetic gradualism,' *Models in Paleobiology*, edited by Schopf, Freeman, Cooper, San Francisco 1972.

Eldredge, Niles, and Stephen J. Gould, 'Punctuated equilibria: the tempo and mode of evolution reconsidered,' *Paleobiology*, **3** (1977).

Eldredge, Niles,*Unfinished Synthesis. Biological Hierarchies and Modern Evolutionary Thought*, Oxford University Press, Oxford 1985.

Elkin, A.P. *The Australian Aborigines*, Angus & Robertson, Sydney 1942.

Franz, Marie-Louise von, *Psyche and Matter*, Shambhala, Boston and London, 1992.

Fröhlich, H. (ed.), *Biological Coherence and Response to External Stimuli*, Springer Verlag, Heidelberg 1988.

Gabor, Dennis, 'A New Microscopic Principle,' *Nature*, **161** (1946).

Goodwin, Brian, 'Development and evolution,' *Journal of Theoretical Biology*, **97**, 1982.

Goodwin, Brian, 'Organisms and minds as organic forms,' *Leonardo*, **22**, 1, (1989).

Gould, Stephen J., Irrelevance, submission and partnership: the changing role of paleontology in Darwin's three centennials, and a modest proposal for macroevolution,' D. Bendall, (ed.), *Evolution from Molecules to Men*, Cambridge University Press, Cambridge 1983.

Gray, C.M., P. Konig, A.K. Engel, and W. Singer, 'Oscillatory responses in cat visual cortex exhibit inter-columnar synchronization which reflects global stimulus properties.' *Nature* **338** (1989), 334-337.

Grimes, W. and K.J. Aufderheide, *Cellular Aspects of Pattern Formation: The Problem of Assembly*, Karger, New York and Basel, 1991.

Grof, Stanislav, with Hal Zina Bennett, *The Holotropic Mind*. Harper San Francisco, 1993.

Grundler W., and F. Kaiser, 'Experimental evidence for coherent excitations correlated with cell growth.' *Nanobiology*, **1** (1992).

Gunzig,E., J.Geheniau and I. Prigogine, 'Entropy and Cosmology,' *Nature*, **330**, 6149 (December 1987).

Haisch, Bernhard Alfonso Rueda and Harold E. Puthoff, 'Beyond $E = mc^2$' *The Sciences*, November/December 1994.

Haisch, Bernhard, Alfonso Rueda, and H.E. Puthoff, 'Inertia as a zero-point-field Lorentz force,' *Physical Review A*, **49.2** (February 1994).

Hall, B.G. 'Evolution on a petri dish.' *Evolutionary Biology*, **15** (1982).

Hansen, G.M., M. Schlitz and C. Tart, 'Summary of remote viewing research,' in Russell Targ and K. Harary, *The Mind Race*. 1972-1982. Villard, New York 1984.

Harris, Errol E., 'The universe in the light of contemporary scientific developments,' in M. Kafatos (ed.), *Bell's Theorem, Quantum Theory and Conceptions of the Universe*. Kluwer Academic Publishers, New York 1989.

Harris, Errol E., *Cosmos and Anthropos*, Humanities Press, New York 1991.

Heisenberg, Werner, *Physics and Philosophy*, Harper & Row, New York 1985.

Ho, M.W., 'On not holding nature still: evolution by process, not by consequence,' in *Evolutionary Processes and Metaphors*, M.W.Ho and S.W. Fox (eds.), Wiley, London 1988.

Ho, M.W., 'The role of action in evolution,' *Cultural Dynamic* **4** (1991).

Honorton, C., R. Berger, M. Varvoglis, M. Quant., P. Derr, E. Schechter, and D. Ferrari, 'Psi-communication in the Ganzfeld: Experiments with an automated testing system and a comparison with a meta-analysis of earlier studies.' *Journal of Parapsychology*, **54** (1990).

Hoyle, Fred, *The Intelligent Universe*, Michael Joseph, London 1983.

Hoyle, F., G. Burbidge and J.V. Narlikar, 'A Quasi-steady state cosmology model with creation of matter,' *The Astrophysical Journal* **410** (20 June 1993).

Ives, Herbert, 'Extrapolation from the Michelson-Morley experiment,' *Journal of the Optical Society of America*, **40** (1950).

Ives, Herbert, 'Light signals sent around a closed path,' *Journal of the Optical Society of America*, **28** (1938).

Ives, Herbert, 'Lorentz-type transformations as derived from performable rod and clock operations,' *Journal of the Optical Society of America*, **39** (1949).

Ives, Herbert, 'Revisions of the Lorentz transformations,' *Proceedings of the American Philosophical Society*, Vol. 95 (1951);

Jacob, François, *The Logic of Life: A History of Heredity*, Pantheon, New York 1970.

Jahn R.G., and B.J. Dunne, 'On the quantum mechanics of consciousness, with application to anomalous phenomena.' *Foundations of Physics*, **16**, 8 (1986).

Jansen B.H., and M.E. Brandt (eds.), *Nonlinear Dynamical Analysis of the EEG*, World Scientific, London and Singapore 1993.

Jeans, J.H., *Astronomy and Cosmogony*, Cambridge University Press, Cambridge, 1929.

Jung, Carl G., 'Commentary on *The Secret of the Golden Flower*', in R. Wilhelm, *The Secret of the Golden Flower*, Harcourt, Brace & World, New York 1962.

Jung, Carl G.,'Ein Brief zur Frage der Synchronizität,' *Zeitschrift für Parapsychologie und Grenzgebiete der Psychologie*, **1** (1961).

Kafatos, M. (ed.), *Bell's Theorem, Quantum Theory and Conceptions of the Universe*, Kluwer Academic Publishers, New York 1989.

Kauffman, Stuart *The Origins of Order: Self-Organization and Selection in Evolution*. Oxford University Press, Oxford 1993.

Kelly, M.T., M.P. Varvoglis and P. Keane, 'Physiological response during psi and sensory presentation of an arousing stimulus,' *Research in Parapsychology*, .G.Roll (ed.), Scarecrow Press, Metuchen, NJ, 1979.

Kleppner, David, Michael Lettmann and Myron Zimermann, 'Highly excited atoms,' *Scientific American* (May 1981).

Krippner, S., W. Braud, I.L. Child, J. Palmer, K.R. Rao, M. Schlitz, R.A.White, and J. Utts, 'Demonstration research and meta-analysis in parapsychology.' *Journal of Parapsychology*, **57** (1993).

Lashley, Karl, 'The problem of cerebral organization in vision,' Biological Symposia, **VII**, *Visual Mechanisms*, Jacques Cattell Press, Lancaster 1942.

Laszlo, Ervin, *Aux Racines de l'Univers*. Fayard, Paris 1992.

Laszlo, Ervin, *Evolution: The Grand Synthesis*. New Science Library, Shambhala, Boston and London 1987.

Laszlo, Ervin, *Introduction to Systems Philosophy: Toward a New Paradigm of Contemporary Thought*. Gordon and Breach, New York 1972, reprinted 1987.

Laszlo, Ervin, *The Creative Cosmos*. Floris Books, Edinburgh 1994.

Licata, Ignazio, 'Dinamica Reticolare dello Spazio-Tempo' [Reticular dynamics of spacetime], Inediti **27**, Soc. Ed. Andromeda, Bologna 1989.

Lorimer, David, *Whole In One: The Near-Death Experience and the Ethic of Interconnectedness*, Arkana, London 1990.

McKenna, Terence, Rupert Sheldrake and Ralph Abraham, *Trialogues at the Edge of the West*. Bear & Co., Santa Fe, NM 1992.

Michelson, A.A., 'The Relative Motion of the Earth and the Luminiferous Ether,' *American Journal of Science*, **22** (1881).

Millar, B. ,'The observational theories: A primer. *European Journal of Parapsychology*, **2** (1978).

Moody, Jr. Raymond A., *The Light Beyond*, Bantam Books, New York 1988.

Moody, Jr., Raymond A., *Life After Life*, Mockingbird Books, Covington 1975.

Netherton, Morris and Nancy Shiffrin, *Past Lives Therapy*, William Morrow, New York 1978.

Peebles, P.J.E. D.N. Schramm, E.L. Turner and P.G. Kron, 'The case for the relativistic hot Big Bang cosmology,' *Nature*, **352** (29 August 1991).

Persinger, Michael A., and Stanley Krippner, 'Dream ESP experiments and geomagnetic activity,' *The Journal of the American Society for Psychical Research*, **83** (1989).

Pool, R., 'Is it healthy to be chaotic?' *Science*, **243** (1989) 604-607.

Pribram, Karl, *Brain and Perception: Holonomy and Structure in Figural Processing*, The MacEachran Lectures. Lawrence Erlbaum, Hillsdale, NJ, 1991.

Price, Huw, 'A neglected route to realism about quantum mechanics', *Mind*, **103** (July 1994).

Prigogine, I. J. Geheniau, E. Gunzig, and P. Nardone, 'Thermodynamics of Cosmological Matter Creation,' *Proceedings of the National Academy of Sciences, USA*, **85** (1988).

Prigogine, Ilya , 'Why irreversibility? The formulation of classical and quantum mechanics for nonintegrable systems,' *International Journal of Quantum Chemistry* (1994).

Prigogine, Ilya, *Thermodynamics of Irreversible Processes*, Wiley-Interscience, New York 1967 (3rd ed.).

Prigogine, Ilya and Isabelle Stengers, *Order out of Chaos: Man's new dialogue with nature*, Bantam Books, New York 1984.

Puthoff, Harold A., 'Source of vacuum electromagnetic zero-point energy,' *Physical Review* A, **40**.9 (1989).

Rein, Glen, 'Modulation of neurotransmitter function by quantum fields.' *Planetary Association for Clean Energy*, **6**,4 (1993).

Rein, Glen, Biological interactions with scalar energy-cellular mechanisms of action, *Proceedings of the 7th International Association of Psychotronics Research*, Georgia, December 1988.

Requardt, Manfred 'From 'Matter-Energy' to 'Irreducible Information Processing' —Arguments for a Paradigm Shift in Fundamental Physics,' *Evolution of Information Processing Systems*, K. Haefner (ed.), Springer Verlag, New York and Berlin 1992.

Rosenthal, R., 'Combining results of independent studies,' *Psychological Bulletin*, **85** (1978).

Russell, J. Scott, *Report on Waves*, British Association for the Advancement of Science, 1845.

Sagnac, G., 'The luminiferous ether demonstrated by the effect of the relative motion of the ether in an interferometer in uniform rotation,' *Comptes Rendus de l'Académie des Sciences*, Paris, **157** (1913).

Saunders, Peter T., 'Evolution without natural selection,' *Journal of Theoretical Biology* (1993).

Schouten, Sybo A., 'Applied parapsychology studies of psychics and healers,' *Journal of Scientific Exploration*, **7**,4 (1993).

Shaara, L., and A. Strathern, 'A preliminary analysis of the relationship between altered states of consciousness, healing, and social structure.' *American Anthropologist* **94** (1992), 145-160.

Sheldrake, Rupert, *A New Science of Life*, Blond & Briggs, London 1981.

Sheldrake, Rupert, *The Presence of the Past*, Times Books, New York 1988.

Silvertooth, Ernest W., 'A new Michelson-Morley experiment,' *Physics Essays*, **5** (1992).

Silvertooth, Ernest W., 'Experimental detection of the ether,' *Speculations in Science and Technology*, **10** (1987);

Silvertooth, Ernest W., 'Motion through the ether,'*Electronics and Wireless World*, May 1989.

Stanford, R. G., 'Experimental psychokinesis: A review from diverse perspectives,' in B.B. Wolman (ed.), *Handbook of Parapsychology*, Van Nostrand Reinhold, New York 1977.

Stapp, Henry P., *Matter, Mind, and Quantum Mechanics*, Springer Verlag, New York 1993.

Stapp, Henry P., 'Quantum Theory and the Place of Mind in Nature,' in *Niels Bohr and Contemporary Philosophy*, J. Faye and H.J. Folse, (eds.) (forthcoming).

Staune, Jean, 'La révolution quantique et ses conséquences sur notre vision du monde'. *3e Millénaire*, **15** (1989).

Stevenson, Ian, 'Birthmarks and birthdefects corresponding to wounds on deceased persons,' *Journal of Scientific Exploration*, **7**,4 (1993).

Stevenson, Ian, *Children Who Remember Previous Lives*, University Press of Virginia, Charlottesville 1987.

Stevenson, Ian, *Unlearned Lanquage: New Studies in Xenoglossy*, University Press of Virginia, Charlottesville 1984.

Targ, Russell and Harold Puthoff, 'Information transmission under conditions of sensory shielding,' *Nature*, **252**, 5476 (1974).

Targ, Russell, 'Remote viewing replication: evaluated by concept analysis.' Dick J. Bierman (ed.), *Proceedings of Presented Papers*, Parapsychological Association 37th Annual Convention, University of Amsterdam, 1994.

Targ, Russell, 'What I see when I close my eyes,' *Journal of Scientific Exploration*, **8**,1 (1994).

Tart, C.T., 'Physiological correlates of psi cognition,' *International Journal of Parapsychology*, **5** (1963).

Törö, Imre (ed.), *Az Élet Alapjai* (Foundations of Life), Gondolat, Budapest 1989.

Ullman M., and S. Krippner, *Dream Studies and Telepathy: An Experimental Approach*, Parapsychology Foundation, New York 1970.

Varvoglis, Mario, 'Goal-directed- and observer-dependent PK: An evaluation of the conformance-behavior model and the observation theories,' *The Journal of the American Society for Psychical Research*, **80** (1986).

Varvoglis, Mario, 'Guérison psychique: Recherches experimentales et hypothèses théoriques,' *Revue Francaise de Psychotronic*, **2**,3 (1989).

Varvoglis, Mario, 'Nonlocality on a human scale: Psi and consciousness research.' LRIP, Paris 1994 (mimeo).

Walker, E.H., 'Foundations of paraphysical and parapsychological phenomena,' in L. Oteri (ed.), *Quantum Physics and Parapsychology*, Parapsychology Foundation, New York 1975.

Wheeler, John Archibald, 'Bits, quanta, meaning,' in *Problems of Theoretical Physics,* A. Giovannini, F. Mancini, and M. Marinaro (eds.), University of Salerno Press, Salerno 1984.

Wheeler, John Archibald, 'Quantum Cosmology', *World Science*, L.Z. Fang and R. Ruffini (eds.), Singapore 1987.

White, John, *The Meeting of Science and Spirit*, Paragon House, New York 1990.

Whittaker, E.T., 'On the partial differential equations of mathematical physics', *Mathematische Annalen*, **57** (1903), 333-355.

Wigner, Eugene, *The Scientist Speculates*, I.J.Good (ed.), Heinemann, London 1961.

Woolger, Roger, *Other Lives, Other Selves,* Doubleday, New York 1987.

Young, J. Z., 'Memory,' Richard Gregory (ed.), *Oxford Companion to the Mind,* Oxford University Press, UK, 1987.

INDEX OF SUBJECTS

adaptation, 75
algorithms, 5
altered states of consciousness, 102, 131
alternative cosmologies, 63
alternative universes, 52
Anomalous recall, 93
Anthropic Principle, 62
archetypes, 129, 135
attractors, 36, 96
backward causation, 57
bacteria, 75
BB (Big Bang) scenario, 63, 68
bifurcation, 7, 8
Big Bang, 63, 119
bio-PK, 91, 104, 105
bioelectromagnetics, 37
bioenergetics, 37
biofield, 80, 84
biological field, 42
biophoton research, 37
birth of the universe, 122
black holes, 18, 66, 120
Boltzmann constant, 36
brain holo-tester, 103
brain, 88, 95, 97, 101, 115, 129
butterfly effect, 96
Cambrian explosion, 76

Casimir-effect, 18
causes and effects, 11
cell, 74
chance versus necessity, 3
chance, 4, 5, 42, 62, 77, 78
chaos dynamics, 81
chaos, 3, 35, 36, 83, 95, 145
chaotic attractors, 36
charged masses, 32
classical thermodynamics, 6
closed systems, 6
coherence, 9, 53, 59, 81, 133
coherent interactions, 96
collective unconscious, 129, 135
complexity, 27
computer simulation, 9
concept of the quantum, 34
conceptual imagination, v
convergence property, 8
convergence, 4, 9, 44
Copenhagen interpretation, 50, 53
cortex, 82
cosmic background radiation, 20
cosmological concepts, 119
cosmological scenario, 120
cosmology, 60
cosmos, 3

Edelman, Martin 83, 98, 99
Einstein, Albert 10, 16, 17, 19, 25, 26, 50,52, 53, 56, 126, 127, 137, 140, 144
Eldredge, Niles 75, 76
Elkin, A.P. 89
Everett, H. 52
Feynman, Richard 55
Fondi, Roberto xv
Fourier, Jean-Jacques 33, 36, 67, 84, 97
Franz, Marie-Louise von 135
Fresnel, A.F. 16, 17, 19
Gabor, Dennis 12, 101, 145
Gallup, George Jr., 93
Galois, Evariste 131
Gauss, Karl Friedrich 131
Geheniau, J. 65, 66, 120
Gilson, R.J. xv, 77
Goodwin, Brian 42, 80
Gould, Jay 75, 76
Grandpierre, Attila xv, 113
Grof, Stanislav 94, 135
Gunzig, E. 65, 120
Gurwitsch, Alexander 42, 80
Guth, Alan 68
Haisch, Bernhard xv, 19, 21, 29, 30
Hardy, Christine xv
Harman, Willis xv
Harris, Errol xv, 61
Hawking, Stephen 18, 66, 68, 126
Heisenberg, Werner 39, 40, 42, 50, 52, 113, 125
Ho, Mae-Won 73, 78

Honorton, C. 92
Hoyle, Fred 14, 64, 83
Huygens, Constantijn 19
Ives, Herbert 19
Jacob, François 74
Jahn, R.G. 52
James, William 134
Jeans, Sir James Hopwood 63
Josephson, Brian 59, 96, 127
Jung, Carl 135
Kant, Emanuel 125
Katchalsky, Aharon 6
Kauffman, Stuart 9
Keane, P. 104
Kelly, M.T. 104
Keys, Donald xv
Kolchov, N.K. 80
Krippner, Stanley xv, 90, 93
Kübler-Ross, Elisabeth 93
Lamarck, Jean-Baptiste 75
Langevin, Paul 19
Lashley, Karl 97, 98, 100
Laszlo, Ervin v-vii, 121, 122, 143-146
LaViolette, Paul xv
Lavoisier, A.L. 140
Leibniz, Friedrich 145
Lewontin, Richard 79
Licata, Ignazio xv, 17
Lorentz, Hendrik A. 18, 19, 20, 29, 30, 127
Lorimer, David xv, 93, 100
Luo Huisheng, xv
Mach, Ernst 19

INDEX OF NAMES

Maslow, Abraham 134
Masulli, Ignazio xv
Maxwell, James Clerk 10, 50
McCulloch, Warren 143
Michelangelo, Buonarroti 130
Michelson, A.A. 16, 17, 19, 20
Millar, B. 105
Milton, John 131
Min, Jiayin, xv
Minkowski, Hermann 66, 126
Moody, Raymond Jr. 93
Moore, Christopher xv
Morley, E.W. 16, 19, 20
Mozart, Wolfgang Amadeus 130, 131
Naess, Arne v-vii, xv, 134
Nardone, P. 65, 120
Narlikar, J.V. 64
Nash, C.B. 104
Netherton, Morris 94
Neumann, John von 9, 94
Newton, Sir Isaac 10, 19, 126
Onsager, Lars 6
Oyama, Susan 79
Pauli, Wolfgang 144
Peat, David xv
Penrose, Roger 126
Persinger, Michael 93
Planck, Max 26, 28, 34, 35, 50, 59, 66, 140
Plato 130, 134
Podolski, Boris 56
Poincaré, Henri 9, 35, 39, 41, 53, 58, 131, 126, 127
Pool, R. 36

Popper, Sir Karl 26
Preparata, G. 80
Pribram, Karl xv, 100, 101, 143-146
Price, Huw 57
Prigogine, Ilya xv, 6, 7, 8, 35, 65, 66, 120, 145
Puthoff, Harold 19, 20, 21, 29, 30, 89, 90, 104
Ravn, Ib xv
Rein, Glen 96
Requardt, Manfred 17
Rhine, J.B. 89
Rosen, Nathan 56
Rosenthal, R. 92
Rueda, Alfonso 19, 29, 30
Russell, J. Scott 29
Rutherford, Ernest 140
Sági, Mária xv
Sagnac, Georges 19
Sah, Sunita 104
Saunders, Peter xv, 78
Schlitz, Marilyn 91
Schouten, S.A. 104
Schrödinger, Erwin 6, 31, 34, 56, 66, 67, 83
Schutzenberger, M. 78
Sermonti, Giuseppe 7, 8
Shakespeare, William 130
Sheldrake, Rupert 42, 43
Siddharta 134
Silvertooth, Ernest 20, 32
Singh, Karan xv
Sitter, Willem De 66
Socrates 134

Spencer, Herbert 5
Stanford, R.G. 105
Stapp, Henry A. xv, 39, 40
Starobinsky, Alexei 68
Staune, Jean xv
Stengers, Isabelle 7
Stevenson, Ian 102
Targ, Russell 89, 90, 95, 104
Teilhard, Pierre de Chardin 6
Tesla, Nicolas 30
Thom, René 42, 80
Thomson, D'Arcy 42
Tyson, Edward 120

Unruh, William 18, 29
Vaidman, L. 113
Varvoglis, Mario xv, 93, 104, 105
Waddington, Conrad 42, 80, 81
Walker, E.H. 105
Weiss, Paul 42, 80
Weyl, Hermann 42, 67
Wheeler, John Archibald xv, 28, 51, 60
White, John xv, 135
Whitehead, Alfred North 6
Whittaker, E.T. 32
Woolger, Roger 94
Young, J.Z. 98

INDEX OF NAMES

Albert, T. 113
Alembert, Jean Le Rond D' 31
Alexander, Samuel 6
Anaximander, 122
Aristotle, 134
Arp, H.C. 64
Aspect, Alain 56
Bateson, Gregory 134
Bauer, Ervin 80
Bearden, Thomas 18
Bell, John 127, 144
Benor, Daniel 92
Bergson, Henri 5
Blake, William 134
Bohm, David 27, 38, 39, 52, 59, 144
Bohr, Niels 50, 51, 53, 125, 140
Boltzmann, Ludwig von 36
Bråten, Henning xv
Braud, Willlam 91
Brier, Søren xv
Broglie, Louis de 50, 56
Burbidge, G. 64
Byrd, Randolph C. 91, 92

Cayce, Edgar 135
Ceruti, Mauro xv, 119-123,
Chalmers, David xv, 54
Chief Seattle, 133
Clarke, C.J.S. xv
Coleridge, Samuel Taylor 131
Combs, Allan xv
Conforti, Michael xv
Confucius, 134
Curran, P.F. 6
Dalton, John 140
Darwin, Charles 73, 75, 78, 79, 138
Davies, Paul 18, 29
Dean, E.D. 104
Del Giudice, E.G. 59, 80, 81
Delanoy, Deborah 104
Democritus 126, 140
Descartes, René 16
Detlefsen, Thorwald 94
Dirac, Paul 18, 25, 53, 144
Donne, John 134
Dunne, B.J. 52
Eccles, Sir John 97

creation cycles, 65
creative act, 131
creative genius, 129
creativity, 43, 129
critical fluctuations, 7
cultural identification, 133
cultural synchronicity, 135
D'Alembert equation, 31
D'Alembert waves, 31
Davies-Unruh effect, 18
De Sitter universe, 66
decision (detection) event, 40, 53
determinism, 39, 42
determinism, vi
Dirac sea, 19, 25
disciplinary gaps, xiii
dissipative systems, 7
distributed memory, 100
divergence property, 8
divergence, 4, 9, 44
DNA, 74
double-slit experiment, 54
dream ESP, 90
dream telepathy, 93
Einfall, 130
electric gradients, 37
Electroencephalograph (EEG) waves, 89, 102, 103
electromagnetic field, 10
electromagnetic waves, 32
electron, 10, 50, 52, 56
electrostatic scalar potential, 18, 31, 32
'elementary quantum phenomenon', 50

ELF waves, 36
embryogenesis, 81
empathetic resonance, 100
empirical testability, 26
energy density, 28
ensembles of systems, 8
EPR experiment, 56, 127
ether, 16, 29
ether-drag, 16
evolution, 3, 7, 8, 15, 21, 69, 75, 76, 77, 138, 139
evolutionary cycles, 69
evolutionary process, 7, 44
evolutionary trajectories, 4, 36, 82
extrasensory perception (ESP), 88, 103
falsifiability, 26
field, 10, 11, 12, 83, 144, 145
fifth field, 15, 16, 140
Final Anthropic Principle, 61
final causes, 3, 26
fluctuations, 7, 35, 120, 122
formative causation, 42
Fourier transformation (transforms), 33, 36, 53, 67
Gabor transforms, 101
Ganzfeld experiments, 92
genetic variation, 78
genome, 74, 79
geological record, 76
geometrodynamics, 31
germline changes, 75
gestalt character, v
global entropy, 62
grand unified theories, xiii

INDEX OF SUBJECTS

gravitation, 10, 15, 21
gravitational clumping, 68
Guassian constraints, 101
healer, 105
heart, 36
Heisenberg operators, 40
Heisenberg quantum universe, 39, 40, 42
hierarchy, 79
holofield, 25, 39, 82, 83, 136
holographic information field, 27
holographic information, 15
holographic principle, 12
holographic processes, 33
holographic receptor patches, 100
'hypothesis of interactive creativity', 130, 132
homeostasis, 146
implicate order, 38
inertia, 19, 21
'in-formation' of the organism, 79
information carrier waves, 115
information storage, 13
information, 14, 67, 145
information-rich zero-point field, 30, 31, 33
initial-conditiondependence, 12
innovation, xv
input-sensitivity of organism, 81
instability, 3, 7, 65, 66
insulation of the genome, 74
Interaction effects in microphysics, 50
interaction quanta/quantum vacuum: *see* QVI

interactive dynamics, 5, 25
intercessory prayer, 92
interconnected system, 4
interconnecting field, 27
interconnection, vi, 4, 8, 9, 14, 15, 16, 83, 144
interference pattern, 54
irreversibility, 6, 35, 121
irreversible change, 138
isolation of quantum, 35
Josephson effect, 59, 96, 127, 133
Josephson junctions, 96
Lamarckian evolution, 75
Lamb-shift, 18
law of gravitation, 11
laws in science, 5, 11, 39
life-review, 93
light-velocity constant, 20, 32
living systems, 36
longitudinal waves, 32
Lorentz transformations, 18
Markovian chain, 14, 29
mass-density, 32
massless charges, 27
mathematical discoveries, 131
matter particles, 18
matter-creation, 64, 69
measurement, 56
memory, 12, 97, 99, 102
meta-universe, 121
Micro- and macro-scale applications, 34
mind, 3, 4, 88, 129
morphic resonance, 42

morphogenetic field, 42, 43, 80
multicyclic cosmology, 64, 65, 66, 121, 122
multicyclic information transfer, 68
multicyclic QVI scenario, 66, 69
mutations, 75, 77, 78, 79
narrative (in science), 143, 144, 146
natural empathies, 134
natural selection, 78
near-death experience (NDE), 93, 94, 97
necessity, 5
negative energy, 64, 65
negentropy, 6
nervous system, 82
nested hierarchies, 15, 73
neural Darwinism, 98
neuronal connections, 82
neurulation, 81, 82
niche, 75
nondynamic correlation, 57
nonequilibrium systems, 6
nonequilibrium thermodynamics, 6
nonlocality, 56
nuclear fields, 15
order, 4, 9, 26, 69, 145
ordered complexity, 4, 5
ordered diversity, 15
ordered systems, 49
organic generation, 81
origin of zero-point field, 20
panoramic memory, 93
paradigm, 137
parapsychology, 88

Participatory Anthropic Principle, 61
particle-creation, 28
past-life experiences, 97
Pauli's exclusion principle, 57
peripheral isolates, 76
pilot wave, 39
Planck-length, 28
Poincaré resonances, 35
Poincaré systems, 8, 34, 39, 41, 53, 58
Poincaré-section, 58
polarized beams, 55
post-Darwinian view, 79
'pre-established harmony', 4
preferential correlations, 4
Prigoginian dynamics, 8
primary and secondary repertoires, 98
principle of uncertainty, 50, 56
probability fields, 10
probability, 35
prompt, 14
psi-research, 95
psyche, 135
punctuated equilibria, 75, 76
Q-factor (Bohm), 38
quantum chromodynamics, 11
quantum field theories, 10
quantum fluctuations, 68
quantum jump, 40, 41
quantum mechanics (theory, physics), 9, 17, 35, 37, 41, 50, 53, 54, 57
quantum nonlocality, 53
quantum phenomenalism, 52
quantum potentials, 96, 97

quantum vacuum, 16, 18, 25, 27, 28, 52, 67,68, 114, 120, 122, 136
quantum-interpretation decision-tree, 51
quarks, 11
Quasi-Steady State Cosmology, 64
QVI (quantum/vacuum interaction), 49, 59, 60, 82, 88, 103, 121, 134, 136, 144, 146
QVI dynamics in the brain, 113
QVI, vi
random variation, 78
randomness, 4, 7, 14, 57, 59, 121
randomness, vii
realism, 50
reality of particles, 40
recall, 97, 98, 100, 102
redshift parameter, 65
reference frame, universal, 17, 20, 141
regeneration, 81, 83
regression therapy, 94
regulatory circuits, 74
remote viewing, 89
resonance, 43, 97, 116
reversal of time, 57
Robertson-Walker universe, 66
Rubik cube, 14
Rydberg atom, 57, 58
scalar energy, 96
scalar patterns, 34
scalar wave-propagations, 31
scalar waves, 30, 31,. 96, 114
scalar-mediated spectrum, 55
scale-invariant fluctuations, 68
Schrödinger waves, 31, 39

Schrödinger's cat., 56
Schrödinger-hologram, 31, 83
second law of thermodynamics, 62
self-generating cosmological feedback cycle, 21
self-referentiality, 4
sensitivity of living state, 37
sensory data, 88
shamans, 89
significant similarities, xiii
simplicity, v
social empathies, 133
social sciences, 139
sociocultural system, 133
solitons, 29
space, 125, 127, 140
spacetime manifold, 126
special theory of relativity, 17
speciation, 75
species, 76, 77
specificity of retranslation, 33
spontaneous communication, 133, 134
steady state concept, 63
stochastic electrodynamics (SED), 21, 28, 59
Strong Anthropic Principle, 61
structural stability, 80
structure of particles, 27, 39
superconductivity, 58
superfluidity, 58
synapses, 97
synthetic theory (biology), 74, 79
synthetic theory, xiii

system of habits, 43
telepathy, 89
telepsychosomatic effect, 105
temporal dimension of QVI, 34
theory of neuronal group selection, 98
thermal energy, 37
thermodynamic equilibrium, 6
time, 127, 128
time-connections, 11
time-symmetry, 35
transdisciplinary unification, 137
transdisciplinary unified theory, xiv, 44, 87, 140, 141
transformations, 145
transparency, 79
transpersonal memory, 101
transpersonal process, 130
universal constants, 61, 62, 63, 69
universal constants, vi

universal cycles, 65
universal field, 15
universal interconnections, 21, 49
universe, 3, 4, 64, 65, 119, 125, 140
unobservability, 26
vacuum (*also see* quantum vacuum), 16, 18, 19, 21, 65, 83
vacuum holofield, 134
vacuum physics, 60
vacuum postulates, 27
virtual particles, 11
wave function, 51, 59, 113, 115
wave interference, 13, 67
Weak Anthropic Principle, 61
Weismann's doctrine, 73
Weyl spacetime cone, 67
xenoglossy, 94
zero-point energy (ZPE), 17, 18, 21
zero-point field (ZPF), 19, 30, 33, 44, 67, 84